Cambridge Elements

Elements in Emerging Theories and Technologies in Metamaterials
edited by
Tie Jun Cui
Southeast University
John B. Pendry
Imperial College London

METAMATERIALS AND NEGATIVE REFRACTION

Rujiang Li
Zhejiang University
Zuojia Wang
Shandong University
Hongsheng Chen
Zhejiang University

CAMBRIDGE
UNIVERSITY PRESS

CAMBRIDGE
UNIVERSITY PRESS

University Printing House, Cambridge CB2 8BS, United Kingdom

One Liberty Plaza, 20th Floor, New York, NY 10006, USA

477 Williamstown Road, Port Melbourne, VIC 3207, Australia

314–321, 3rd Floor, Plot 3, Splendor Forum, Jasola District Centre, New Delhi – 110025, India

79 Anson Road, #06–04/06, Singapore 079906

Cambridge University Press is part of the University of Cambridge.

It furthers the University's mission by disseminating knowledge in the pursuit of education, learning, and research at the highest international levels of excellence.

www.cambridge.org
Information on this title: www.cambridge.org/9781108749237
DOI: 10.1017/9781108782371

First published 2020

A catalogue record for this publication is available from the British Library.

ISBN 978-1-108-74923-7 Paperback
ISSN 2399-7486 (online)
ISSN 2514-3875 (print)

Metamaterials and Negative Refraction

Elements in Emerging Theories and Technologies in Metamaterials

DOI: 10.1017/9781108782371
First published online: November 2020

Rujiang Li
Zhejiang University
Zuojia Wang
Shandong University
Hongsheng Chen
Zhejiang University

Author for correspondence: Hongsheng Chen, hansomchen@zju.edu.cn

Abstract: The discovery of artificial electromagnetic materials, called metamaterials, not only redefines the human perception of constitutive parameters in electromagnetic theory but also brings forward new phenomena, such as negative refraction. We provide a comprehensive introduction to the unique characteristics of metamaterials, starting with Maxwell's equations and the kDB coordinate system and moving through to theoretical concepts of negative refraction in metamaterials. For each kind of media, including isotropic, anisotropic and bianisotropic metamaterials, we discuss the characteristic waves and their properties. We show examples of negative refraction both theoretically and experimentally.

Keywords: metamaterials, negative refraction, kDB coordinate system

ISBNs: 9781108749237 (PB), 9781108782371 (OC)
ISSNs: 2399-7486 (online), 2514-3875 (print)

Contents

1 Introduction

The refraction of electromagnetic waves is a common phenomenon. Nearly all the dielectric media in nature have positive refractive indices. According to Snell's law, the refraction angle for an electromagnetic wave propagating from one medium to another has the same sign as the incident angle, namely the refraction is positive. However, the discovery of artificial electromagnetic materials called metamaterials redefines the perception of constitutive parameters in electromagnetic theory. For a metamaterial with both negative permittivity and negative permeability, its refractive index is negative. When a wave propagates from a natural dielectric medium to a metamaterial with a negative refractive index, the refraction angle has a different sign from the incident angle. We call this phenomenon negative refraction. In this Element, we will introduce several typical kinds of metamaterials with different constitutive tensors and the experimental implementations and applications of these media.

Section 2 is an overview of the basic concepts and physical background that are necessary to study negative refraction. We start from the Maxwell equations and constitutive relations to give a general classification of the electromagnetic media. Then we introduce the waves in the *xyz* coordinate system and their dispersion relations. Specifically, we briefly review the *kDB* coordinate system, which not only simplifies the calculation of waves in complicated media but also gives a clear and intuitive physical insight. Readers can directly touch the intrinsic wave types and wave propagation using the *kDB* coordinate system without caring about the actual *xyz* coordinate system, which is related to the orientations of the media. The subsequent sections are developed based on the *kDB* coordinate system.

From Section 3 onward we start to introduce several kinds of metamaterials with different constitutive tensors and the negative refraction in these media. In Section 3 we discuss the theoretical concepts and design principles of negative refraction in isotropic metamaterials. This simplest case can give readers a basic understanding of negative refraction. In Section 4 we discuss the negative refraction in anisotropic metamaterials, where the uniaxial metamaterials are taken as a typical example. In Section 5 we discuss the negative refraction in bianisotropic metamaterials by taking the biisotropic chiral metamaterials as an example. In each section we analyze the type-I and type-II characteristic waves and their propagation behaviors in conventional materials and in metamaterials. We also show that negative refraction can occur at the interface between the air and metamaterial using the *k* surfaces.

In Section 6 we introduce the experimental implementations of various kinds of metamaterials. According to the fabrication difficulties, we first introduce the implementation of anisotropic metamaterials, then the implementations of bianisotropic metamaterials, chiral metamaterials and isotropic metamaterials. The fabrication of isotropic metamaterials is much more difficult due to the requirement of isotropy.

Section 7 is the concluding remarks. We summarize the structure of the Element and point out the key results. We hope this Element will provide a guidance to undergraduates and graduate students who study the theoretical and experimental aspects of metamaterials, especially the realization of negative refraction in metamaterials.

2 Constitutive Parameters and Negative Refraction

2.1 Maxwell Equations and Constitutive Parameters

The electromagnetic phenomena inside various media are governed by the Maxwell equations, which were established by James Clerk Maxwell in 1864 [1]. In three-dimensional vector notation, the Maxwell equations are

$$\nabla \times \overline{E}(\overline{r},t) = -\frac{\partial}{\partial t}\overline{B}(\overline{r},t), \tag{2.1a}$$

$$\nabla \times \overline{H}(\overline{r},t) = \overline{J}(\overline{r},t) + \frac{\partial}{\partial t}\overline{D}(\overline{r},t), \tag{2.1b}$$

$$\nabla \cdot \overline{D}(\overline{r},t) = \rho(\overline{r},t), \tag{2.1c}$$

$$\nabla \cdot \overline{B}(\overline{r},t) = 0, \tag{2.1d}$$

where $\overline{E}(\overline{r},t)$ is the electric field strength, $\overline{B}(\overline{r},t)$ is the magnetic flux density, $\overline{H}(\overline{r},t)$ is the magnetic field strength, $\overline{D}(\overline{r},t)$ is the electric displacement, $\overline{J}(\overline{r},t)$ is the electric current density and $\rho(\overline{r},t)$ is the electric charge density. Note that all the physical quantities are dependent on both the position $\overline{r}$ and time t. For a time harmonic field with the term $\exp(-i\omega t)$, Eqs. (2.1a, 2.1b, 2.1c, 2.1d) reduce to

$$\nabla \times \overline{E}(\overline{r}) = i\omega\overline{B}(\overline{r}), \tag{2.2a}$$

$$\nabla \times \overline{H}(\overline{r}) = \overline{J}(\overline{r}) - i\omega\overline{D}(\overline{r}), \tag{2.2b}$$

$$\nabla \cdot \overline{D}(\overline{r}) = \rho(\overline{r}), \tag{2.2c}$$

$$\nabla \cdot \overline{B}(\overline{r}) = 0, \tag{2.2d}$$

and the constitutive relations are

$$\overline{D}(\overline{r}) = \overline{\overline{\varepsilon}}(\omega)\cdot\overline{E}(\overline{r}) + \overline{\overline{\xi}}(\omega)\cdot\overline{H}(\overline{r}), \tag{2.3a}$$

$$\overline{B}(\overline{r}) = \overline{\overline{\zeta}}(\omega)\cdot\overline{E}(\overline{r}) + \overline{\overline{\mu}}(\omega)\cdot\overline{H}(\overline{r}), \tag{2.3b}$$

where $\overline{\overline{\varepsilon}}(\omega)$, $\overline{\overline{\xi}}(\omega)$, $\overline{\overline{\zeta}}(\omega)$ and $\overline{\overline{\mu}}(\omega)$ are the constitutive parameters. Since the electromagnetic flux densities $\overline{D}$ and $\overline{B}$ are expressed using the electromagnetic field densities $\overline{E}$ and $\overline{H}$, we call this kind of expression "constitutive relations in the $\overline{EH}$ representation."

At a microscopic level, the electromagnetic properties of materials are understood by the interactions of electromagnetic waves with the charged particles in the composite atoms or molecules. When subjected to electromagnetic fields, the motion of charged particles produces currents and modifies the electromagnetic wave propagation in these media compared to that in free space. By accounting for the presence and behavior of these charged particles from a macroscopic scale, the material medium is characterized by the constitutive relations, which relate the electromagnetic flux densities $\overline{D}$ and $\overline{B}$ with the electromagnetic field densities $\overline{E}$ and $\overline{H}$, as shown in Eqs. (2.3a, 2.3b). "It is worth noting that, since all the microscopic dimensions such as the unit cell dimensions or the mean free paths of electrons are substantially less than a typical wavelength [2], the dependences between the constitutive parameters and the wave vectors of the electromagnetic fields are neglected in this Element.".

According to the constitutive parameters, the electromagnetic media can be mainly divided into four categories: *isotropic* media, *anisotropic* media, *biisotropic* media and *bianisotropic* media. The isotropic media have the simplest constitutive parameters, and this kind of media has been well studied. The bianisotropic media are the most general but complicated media. For the constitutive parameters of all the media, their real and imaginary parts should follow the causality requirement, namely the Kramers-Kronig relations.

Isotropic media are the simplest media, and they widely exist in nature. In isotropic media $\overline{\overline{\xi}} = \overline{\overline{\zeta}} = 0$ and the constitutive relations from Eqs. (2.3a, 2.3b) reduce to

$$\overline{D} = \varepsilon \overline{E}, \tag{2.4a}$$

$$\overline{B} = \mu \overline{H}, \tag{2.4b}$$

where the two constitutive parameters have only one respective matrix element and there are only two independent elements in total. Usually the matrix elements of ε and μ are complex and frequency dependent. For most isotropic materials, the real parts of the electric permittivity ε and the magnetic permeability μ are positive, but there are exceptions. For example, in plasmas the combination of ordinary displacement current density with electron-convection current density can yield a net negative real part of the permittivity for sufficiently low frequencies. When only one of the real parts of ε and μ is negative, plane waves decay exponentially. However, when both the real parts of ε and μ are negative, waves can still propagate in such media.

In anisotropic media $\overline{\overline{\xi}} = \overline{\overline{\zeta}} = 0$ and the constitutive relations from Eqs. (2.3a, 2.3b) reduce to

$$\overline{D} = \overline{\overline{\varepsilon}} \cdot \overline{E}, \tag{2.5a}$$

$$\overline{B} = \overline{\overline{\mu}} \cdot \overline{H}, \tag{2.5b}$$

where each constitutive parameter is a matrix with nine elements and there are 18 independent matrix elements in total. Usually the matrix elements of $\overline{\overline{\varepsilon}}$ and $\overline{\overline{\mu}}$ are complex and frequency dependent. Uniaxial media are a special kind of anisotropic media, where $\overline{\overline{\varepsilon}} = diag\,(\varepsilon, \varepsilon, \varepsilon_z)$, $\overline{\overline{\mu}} = diag\,(\mu, \mu, \mu_z)$ and the two constitutive parameters have two independent matrix elements, respectively. In nature calcite is a typical example of anisotropic media.

In biisotropic media the constitutive relations from Eqs. (2.3a, 2.3b) reduce to

$$\overline{D} = \varepsilon\overline{E} + \xi\overline{H}, \tag{2.6a}$$

$$\overline{B} = \zeta\overline{E} + \mu\overline{H}, \tag{2.6b}$$

where each constitutive parameter has only one element and there are four complex and frequency-dependent matrix elements in total. Usually, the matrix elements of ε, ξ, ζ and μ are complex and frequency dependent. Specifically, if both ξ and ζ are imaginary and the biisotropic media are lossless, the biisotropic media further reduce to the chiral media by letting $\xi \rightarrow i\xi$. The constitutive relations of the chiral media are

$$\overline{D} = \varepsilon\overline{E} + i\xi\overline{H}, \tag{2.7a}$$

$$\overline{B} = -i\xi\overline{E} + \mu\overline{H}, \tag{2.7b}$$

where the condition of lossless media requires that $\zeta = -i\xi$. In this case the four complex matrix elements reduce to three real matrix elements. Since the constitutive parameters are all in isotropic forms, this kind of chiral media can be called biisotropic chiral media.

Bianisotropic media are the most complicated media. In bianisotropic media the constitutive relations are exactly of the general form shown in Eqs. (2.3a, 2.3b). Each constitutive parameter is a matrix with nine elements, and there are 36 independent matrix elements in total. The matrix elements $\overline{\overline{\varepsilon}}$, $\overline{\overline{\xi}}$, $\overline{\overline{\zeta}}$ and $\overline{\overline{\mu}}$ are usually complex and frequency dependent. Specifically, if both $\overline{\overline{\xi}}$ and $\overline{\overline{\zeta}}$ are imaginary and the bianisotropic media are lossless, the bianisotropic media reduce to the chiral media by letting $\overline{\overline{\xi}} \rightarrow i\overline{\overline{\xi}}$. The constitutive relations of the chiral media are

$$\overline{D} = \overline{\overline{\varepsilon}} \cdot \overline{E} + i\overline{\overline{\xi}} \cdot \overline{H}, \tag{2.8a}$$

$$\overline{B} = -i\overline{\overline{\xi}} \cdot \overline{E} + \overline{\overline{\mu}} \cdot \overline{H}, \tag{2.8b}$$

where the condition of lossless media requires that $\overline{\overline{\zeta}} = -i\overline{\overline{\xi}}$. In this case, the 36 complex matrix elements reduce to 27 real matrix elements. Since the constitutive parameters are all in anisotropic form, this kind of chiral media can be called bianisotropic chiral media. There are a few examples of bianistropic media in nature. A material isotropic on its own stationary reference frame exhibits bianisotropic properties when it is observed from a moving frame. The antiferromagnetic material chromium oxide (Cr_2O_3) is a kind of bianisotropic material in its own frame due to the magnetoelectric effects [3].

2.2 Waves in the *xyz* Coordinate System

We focus on the Maxwell equations in a linear, homogeneous and source-free medium. For a time harmonic field with the term $\exp(-i\omega t)$, Maxwell equations in the *xyz* coordinate system are

$$\nabla \times \overline{E}(\overline{r}) = i\omega \overline{B}(\overline{r}), \tag{2.9a}$$

$$\nabla \times \overline{H}(\overline{r}) = -i\omega \overline{D}(\overline{r}), \tag{2.9b}$$

$$\nabla \cdot \overline{D}(\overline{r}) = 0, \tag{2.9c}$$

$$\nabla \cdot \overline{B}(\overline{r}) = 0, \tag{2.9d}$$

with the constitutive relations the same as Eqs. (2.3a, 2.3b) in the $\overline{EH}$ representation. Further, for a plane wave solution with the term $\exp\left(i\overline{k} \cdot \overline{r}\right)$, the Maxwell equations reduce to

$$\overline{k} \times \overline{E} = \omega \overline{B}, \tag{2.10a}$$

$$\overline{k} \times \overline{H} = -\omega \overline{D}, \tag{2.10b}$$

$$\overline{k} \cdot \overline{D} = 0, \tag{2.10c}$$

$$\overline{k} \cdot \overline{B} = 0, \tag{2.10d}$$

where $\overline{k}$ is the wave vector.

In order to find the dispersion relations, we start from Eqs. (2.10a, 2.10b) and replace $\overline{D}$ and $\overline{B}$ with $\overline{E}$ and $\overline{H}$, respectively. The results are

$$\overline{k} \times \overline{E} - \omega \overline{\overline{\zeta}} \cdot \overline{E} = \omega \overline{\overline{\mu}} \cdot \overline{H}, \tag{2.11a}$$

$$\overline{k} \times \overline{H} + \omega \overline{\overline{\xi}} \cdot \overline{H} = -\omega \overline{\overline{\varepsilon}} \cdot \overline{E}, \tag{2.11b}$$

Alternatively, we can write in the following matrix forms

$$\begin{pmatrix} -\zeta_{11} & -v_z-\zeta_{12} & v_y-\zeta_{13} \\ v_z-\zeta_{21} & -\zeta_{22} & -v_x-\zeta_{23} \\ -v_y-\zeta_{31} & v_x-\zeta_{32} & -\zeta_{33} \end{pmatrix}\begin{pmatrix} E_x \\ E_y \\ E_z \end{pmatrix} = \begin{pmatrix} \mu_{11} & \mu_{12} & \mu_{13} \\ \mu_{21} & \mu_{22} & \mu_{23} \\ \mu_{31} & \mu_{32} & \mu_{33} \end{pmatrix}\begin{pmatrix} H_x \\ H_y \\ H_z \end{pmatrix}, \tag{2.12a}$$

$$\begin{pmatrix} -\xi_{11} & v_z-\xi_{12} & -v_y-\xi_{13} \\ -v_z-\xi_{21} & -\xi_{22} & v_x-\xi_{23} \\ v_y-\xi_{31} & -v_x-\xi_{32} & -\xi_{33} \end{pmatrix}\begin{pmatrix} H_x \\ H_y \\ H_z \end{pmatrix} = \begin{pmatrix} \varepsilon_{11} & \varepsilon_{12} & \varepsilon_{13} \\ \varepsilon_{21} & \varepsilon_{22} & \varepsilon_{23} \\ \varepsilon_{31} & \varepsilon_{32} & \varepsilon_{33} \end{pmatrix}\begin{pmatrix} E_x \\ E_y \\ E_z \end{pmatrix}. \tag{2.12b}$$

where $\overline{v} = \overline{k}/\omega$. For waves propagating in a linear, homogeneous and source-free medium, Eqs. (2.12a, 2.12b) should be fulfilled. After the two equations are combined, the determinant of the coefficient matrix should be zero, namely

$$\left| -i\overline{\overline{V}} - \overline{\overline{\zeta}} - \overline{\overline{\mu}} \cdot \left(i\overline{\overline{V}} - \overline{\overline{\xi}} \right)^{-1} \cdot \overline{\overline{\varepsilon}} \right| = 0, \tag{2.13}$$

where $\overline{\overline{V}} = v_x\overline{\overline{T}}_1 + v_y\overline{\overline{T}}_2 + v_z\overline{\overline{T}}_3$ and

$$\overline{\overline{T}}_1 = \begin{pmatrix} 0 & 0 & 0 \\ 0 & 0 & -i \\ 0 & i & 0 \end{pmatrix}, \quad \overline{\overline{T}}_2 = \begin{pmatrix} 0 & 0 & i \\ 0 & 0 & 0 \\ -i & 0 & 0 \end{pmatrix}, \quad \overline{\overline{T}}_3 = \begin{pmatrix} 0 & -i & 0 \\ i & 0 & 0 \\ 0 & 0 & 0 \end{pmatrix}. \tag{2.14}$$

2.3 Waves in the *kDB* Coordinate System

The calculation of wave propagation in the *xyz* coordinate system is complicated, especially for the anisotropic media or even more complicated media, since generally all three vectors, $\overline{k}$, $\overline{D}(\overline{E})$ and $\overline{B}(\overline{H})$, have three components, respectively. To simplify this issue, in this Element we will discuss the wave propagation and demonstrate the negative refraction in the *kDB* coordinate system. The *kDB* coordinate system is proposed by J. A. Kong [4]. We will show in the following sections that use of the *kDB* coordinate system can simplify the calculations and, more importantly, give clear and intuitive physical insight. This section is a brief introduction of the general calculation procedure. For details, the reader can refer to [4].

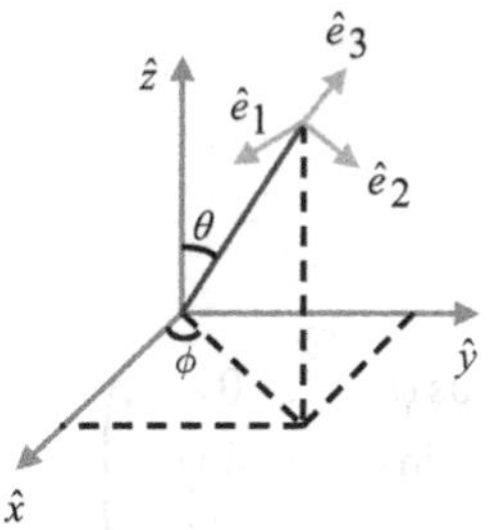

Figure 2.1 A schematic of the *kDB* coordinate system.

2.3.1 Coordinate Transformation

Here we focus on the Maxwell equations in a linear, homogeneous and source-free medium. From Eqs. (2.10a, 2.10b, 2.10c, 2.10d), we can see that the wave vector $\overline{k}$ is always perpendicular to both the electric displacement $\overline{D}$ and the magnetic flux density $\overline{B}$. But $\overline{D}$ and $\overline{B}$ may not be perpendicular to each other. We define the plane formed by $\overline{D}$ and $\overline{B}$ and perpendicular to $\overline{k}$ as the $\overline{DB}$ plane. Meanwhile, we construct a *kDB* coordinate system, which is formed by the three vectors $\overline{k}$, $\overline{D}$ and $\overline{B}$.

The *kDB* coordinate system is constructed with help from the spherical coordinate system. As shown in Fig. 2.1, the unit vectors $\hat{e}_1$, $\hat{e}_2$ and $\hat{e}_3$ are defined as

$$\hat{e}_1 = -\hat{\phi} = \hat{x}\sin\phi - \hat{y}\cos\phi, \tag{2.15a}$$

$$\hat{e}_2 = \hat{\theta} = \hat{x}\cos\theta\cos\phi + \hat{y}\cos\theta\sin\phi - \hat{z}\sin\theta, \tag{2.15b}$$

$$\hat{e}_3 = \hat{r} = \hat{x}\sin\theta\cos\phi + \hat{y}\sin\theta\sin\phi + \hat{z}\cos\theta, \tag{2.15c}$$

where the three unit vectors are orthogonal to each other. In this coordinate system $\overline{k}$ is in the direction of $\hat{e}_3$, namely $\overline{k} = \hat{e}_3 k = \hat{x}k\sin\theta\cos\phi + \hat{y}k\sin\theta\sin\phi + \hat{z}k\cos\theta$. $\overline{D}$ and $\overline{B}$ are in the plane formed by $\hat{e}_1$ and $\hat{e}_2$, which can be expressed as $\overline{D} = \hat{e}_1 D_1 + \hat{e}_2 D_2$ and $\overline{B} = \hat{e}_1 B_1 + \hat{e}_2 B_2$. In the following, we will show the transformation between the *xyz* coordinate system and the *kDB* coordinate system for both the fields and the constitutive parameters.

First, we consider a vector, $\overline{A}$, which has the form $\overline{A} = \hat{x}A_x + \hat{y}A_y + \hat{z}A_z$ in the *xyz* coordinate system, while in the *kDB* coordinate system, its form is $\overline{A}_k = \hat{e}_1 A_1 + \hat{e}_2 A_2 + \hat{e}_3 A_3$. Since the two forms correspond to the same vector, the relations between the components are

$$A_1 = \hat{e}_1 \cdot \overline{A}_k = \hat{e}_1 \cdot \overline{A} = \sin\phi A_x - \cos\phi A_y, \tag{2.16a}$$

$$A_2 = \hat{e}_2 \cdot \overline{A}_k = \hat{e}_2 \cdot \overline{A} = \cos\theta\cos\phi A_x + \cos\theta\sin\phi A_y - \sin\theta A_z, \tag{2.16b}$$

$$A_3 = \hat{e}_3 \cdot \overline{A}_k = \hat{e}_3 \cdot \overline{A} = \sin\theta\cos\phi A_x + \sin\theta\sin\phi A_y + \cos\theta A_z, \tag{2.16c}$$

and the matrix form is

$$\overline{A}_k = \overline{\overline{T}} \cdot \overline{A}, \tag{2.17}$$

where

$$\overline{\overline{T}} = \begin{pmatrix} \sin\phi & -\cos\phi & 0 \\ \cos\theta\cos\phi & \cos\theta\sin\phi & -\sin\theta \\ \sin\theta\cos\phi & \sin\theta\sin\phi & \cos\theta \end{pmatrix} \tag{2.18}$$

is the transformation matrix. Conversely, the transformation from the *kDB* coordinate system to the *xyz* coordinate system is

$$\overline{A} = \overline{\overline{T}}^{-1} \cdot \overline{A}_k, \tag{2.19}$$

where

$$\overline{\overline{T}}^{-1} = \overline{\overline{T}}^{T} = \begin{pmatrix} \sin\phi & \cos\theta\cos\phi & \sin\theta\cos\phi \\ -\cos\phi & \cos\theta\sin\phi & \sin\theta\sin\phi \\ 0 & -\sin\theta & \cos\theta \end{pmatrix}. \tag{2.20}$$

The fields $\overline{D}$, $\overline{B}$, $\overline{E}$ and $\overline{H}$ in the Maxwell equations are all vectors. Thus their transformations must follow the same rule. We directly have

$$\overline{D}_k = \overline{\overline{T}} \cdot \overline{D}, \tag{2.21a}$$

$$\overline{B}_k = \overline{\overline{T}} \cdot \overline{B}, \tag{2.21b}$$

$$\overline{E}_k = \overline{\overline{T}} \cdot \overline{E}, \tag{2.21c}$$

$$\overline{H}_k = \overline{\overline{T}} \cdot \overline{H}, \tag{2.21d}$$

and

$$\overline{D} = \overline{\overline{T}}^{-1} \cdot \overline{D}_k, \tag{2.22a}$$

$$\overline{B} = \overline{\overline{T}}^{-1} \cdot \overline{B}_k, \tag{2.22b}$$

$$\overline{E} = \overline{\overline{T}}^{-1} \cdot \overline{E}_k, \tag{2.22c}$$

$$\overline{H} = \overline{\overline{T}}^{-1} \cdot \overline{H}_k. \tag{2.22d}$$

Besides, the wave vectors also have the transformation relations

$$\overline{k}_k = \overline{\overline{T}} \cdot \overline{k}, \tag{2.23a}$$

$$\overline{k} = \overline{\overline{T}}^{-1} \cdot \overline{k}_k. \tag{2.23b}$$

Next, we need to calculate the transformation of the constitutive parameters, which are generally second-order tensors. Usually the constitutive relations are expressed in the $\overline{EH}$ representation as

$$\overline{D} = \overline{\overline{\varepsilon}} \cdot \overline{E} + \overline{\overline{\xi}} \cdot \overline{H}, \tag{2.24a}$$

$$\overline{B} = \overline{\overline{\zeta}} \cdot \overline{E} + \overline{\overline{\mu}} \cdot \overline{H}. \tag{2.24b}$$

For convenience here we express the constitutive relations in the $\overline{DB}$ representation. After calculations the relations are

$$\overline{E} = \overline{\overline{\kappa}} \cdot \overline{D} + \overline{\overline{\chi}} \cdot \overline{B}, \tag{2.25a}$$

$$\overline{H} = \overline{\overline{\gamma}} \cdot \overline{D} + \overline{\overline{\nu}} \cdot \overline{B}, \tag{2.25b}$$

where

$$\overline{\overline{\kappa}} = \left(\overline{\overline{\xi}}^{-1} \cdot \overline{\overline{\varepsilon}} - \overline{\overline{\mu}}^{-1} \cdot \overline{\overline{\zeta}}\right)^{-1} \cdot \overline{\overline{\xi}}^{-1}, \tag{2.26a}$$

$$\overline{\overline{\chi}} = \left(\overline{\overline{\mu}}^{-1} \cdot \overline{\overline{\zeta}} - \overline{\overline{\xi}}^{-1} \cdot \overline{\overline{\varepsilon}}\right)^{-1} \cdot \overline{\overline{\mu}}^{-1}, \tag{2.26b}$$

$$\overline{\overline{\gamma}} = \left(\overline{\overline{\varepsilon}}^{-1} \cdot \overline{\overline{\xi}} - \overline{\overline{\zeta}}^{-1} \cdot \overline{\overline{\mu}}\right)^{-1} \cdot \overline{\overline{\varepsilon}}^{-1}, \tag{2.26c}$$

$$\overline{\overline{\nu}} = \left(\overline{\overline{\zeta}}^{-1} \cdot \overline{\overline{\mu}} - \overline{\overline{\varepsilon}}^{-1} \cdot \overline{\overline{\xi}}\right)^{-1} \cdot \overline{\overline{\zeta}}^{-1}, \tag{2.26d}$$

and

$$\overline{\overline{\varepsilon}} = \left(\overline{\overline{\chi}}^{-1} \cdot \overline{\overline{\kappa}} - \overline{\overline{\nu}}^{-1} \cdot \overline{\overline{\gamma}}\right)^{-1} \cdot \overline{\overline{\chi}}^{-1}, \tag{2.27a}$$

$$\overline{\overline{\xi}} = \left(\overline{\overline{\nu}}^{-1} \cdot \overline{\overline{\gamma}} - \overline{\overline{\chi}}^{-1} \cdot \overline{\overline{\kappa}}\right)^{-1} \cdot \overline{\overline{\nu}}^{-1}, \tag{2.27b}$$

$$\overline{\overline{\zeta}} = \left(\overline{\overline{\kappa}}^{-1} \cdot \overline{\overline{\chi}} - \overline{\overline{\gamma}}^{-1} \cdot \overline{\overline{\nu}}\right)^{-1} \cdot \overline{\overline{\kappa}}^{-1}, \tag{2.27c}$$

$$\overline{\overline{\mu}} = \left(\overline{\overline{\gamma}}^{-1} \cdot \overline{\overline{\nu}} - \overline{\overline{\kappa}}^{-1} \cdot \overline{\overline{\chi}}\right)^{-1} \cdot \overline{\overline{\gamma}}^{-1}. \tag{2.27d}$$

Thus, transforming the constitutive relations from the *xyz* coordinate system to the *kDB* coordinate system, we get

$$\overline{E}_k = \overline{\overline{\kappa}}_k \cdot \overline{D}_k + \overline{\overline{\chi}}_k \cdot \overline{B}_k, \tag{2.28a}$$

$$\overline{H}_k = \overline{\overline{\gamma}}_k \cdot \overline{D}_k + \overline{\overline{\nu}}_k \cdot \overline{B}_k, \tag{2.28b}$$

where

$$\overline{\overline{\kappa}}_k = \overline{\overline{T}} \cdot \overline{\overline{\kappa}} \cdot \overline{\overline{T}}^{-1}, \tag{2.29a}$$

$$\overline{\overline{\chi}}_k = \overline{\overline{T}} \cdot \overline{\overline{\chi}} \cdot \overline{\overline{T}}^{-1}, \tag{2.29b}$$

$$\overline{\overline{\gamma}}_k = \overline{\overline{T}} \cdot \overline{\overline{\gamma}} \cdot \overline{\overline{T}}^{-1}, \tag{2.29c}$$

$$\overline{\overline{\nu}}_k = \overline{\overline{T}} \cdot \overline{\overline{\nu}} \cdot \overline{\overline{T}}^{-1}. \tag{2.29d}$$

2.3.2 Maxwell Equations in the *kDB* Coordinate System

If we use the transformation relations in Eqs. (2.22a, 2.22b, 2.22c, 2.22d) and Eq. (2.23b), the Maxwell equations reduce to

$$\overline{\overline{T}} \cdot \left[\left(\overline{\overline{T}}^{-1} \cdot \overline{k}_k \right) \times \left(\overline{\overline{T}}^{-1} \cdot \overline{E}_k \right) \right] = \omega \overline{B}_k, \tag{2.30a}$$

$$\overline{\overline{T}} \cdot \left[\left(\overline{\overline{T}}^{-1} \cdot \overline{k}_k \right) \times \left(\overline{\overline{T}}^{-1} \cdot \overline{H}_k \right) \right] = \omega \overline{D}_k, \tag{2.30b}$$

$$\left(\overline{\overline{T}}^{-1} \cdot \overline{k}_k \right) \cdot \left(\overline{\overline{T}}^{-1} \cdot \overline{D}_k \right) = 0, \tag{2.30c}$$

$$\left(\overline{\overline{T}}^{-1} \cdot \overline{k}_k \right) \cdot \left(\overline{\overline{T}}^{-1} \cdot \overline{B}_k \right) = 0. \tag{2.30d}$$

Note that in the *kDB* coordinate system, $\overline{k}_k$ has only one component, and $\overline{D}_k$ and $\overline{B}_k$ have two components, respectively. Considering Eqs. (2.18, 2.20), we can verify that

$$\overline{\overline{T}} \cdot \left[\left(\overline{\overline{T}}^{-1} \cdot \overline{k}_k \right) \times \left(\overline{\overline{T}}^{-1} \cdot \overline{E}_k \right) \right] = \overline{k}_k \times \overline{E}_k, \tag{2.31a}$$

$$\overline{\overline{T}} \cdot \left[\left(\overline{\overline{T}}^{-1} \cdot \overline{k}_k \right) \times \left(\overline{\overline{T}}^{-1} \cdot \overline{H}_k \right) \right] = \overline{k}_k \times \overline{H}_k, \tag{2.31b}$$

$$\left(\overline{\overline{T}}^{-1} \cdot \overline{k}_k \right) \cdot \left(\overline{\overline{T}}^{-1} \cdot \overline{D}_k \right) = \overline{k}_k \cdot \overline{D}_k, \tag{2.31c}$$

$$\left(\overline{\overline{T}}^{-1} \cdot \overline{k}_k \right) \cdot \left(\overline{\overline{T}}^{-1} \cdot \overline{B}_k \right) = \overline{k}_k \cdot \overline{B}_k. \tag{2.31d}$$

Thus the Maxwell equations in the *kDB* coordinate system are

$$\overline{k}_k \times \overline{E}_k = \omega \overline{B}_k, \tag{2.32a}$$

$$\overline{k}_k \times \overline{H}_k = -\omega \overline{D}_k, \tag{2.32b}$$

$$\overline{k}_k \cdot \overline{D}_k = 0, \tag{2.32c}$$

$$\overline{k}_k \cdot \overline{B}_k = 0, \tag{2.32d}$$

and the constitutive relations are shown in Eqs. (2.28a, 2.28b).

In order to find the dispersion relations, we start from Eqs.(2.32a, 2.32b) and replace $\overline{E}_k$ and $\overline{H}_k$ with $\overline{D}_k$ and $\overline{B}_k$, respectively. The results are

$$\overline{k}_k \times (\overline{\kappa}_k \cdot \overline{D}_k) = \omega \overline{B}_k - \overline{k}_k \times (\overline{\chi}_k \cdot \overline{B}_k), \quad (2.33a)$$

$$\overline{k}_k \times (\overline{\nu}_k \cdot \overline{B}_k) = -\omega \overline{D}_k - \overline{k}_k \times (\overline{\gamma}_k \cdot \overline{D}_k). \quad (2.33b)$$

Writing in the matrix forms, we obtain

$$\begin{pmatrix} \kappa_{11} & \kappa_{12} \\ \kappa_{21} & \kappa_{22} \end{pmatrix} \begin{pmatrix} D_1 \\ D_2 \end{pmatrix} = -\begin{pmatrix} \chi_{11} & \chi_{12} - u \\ \chi_{21} + u & \chi_{22} \end{pmatrix} \begin{pmatrix} B_1 \\ B_2 \end{pmatrix}, \quad (2.34a)$$

$$\begin{pmatrix} \nu_{11} & \nu_{12} \\ \nu_{21} & \nu_{22} \end{pmatrix} \begin{pmatrix} B_1 \\ B_2 \end{pmatrix} = -\begin{pmatrix} \gamma_{11} & \gamma_{12} + u \\ \gamma_{21} - u & \gamma_{22} \end{pmatrix} \begin{pmatrix} D_1 \\ D_2 \end{pmatrix}, \quad (2.34b)$$

where $u = \omega/k$. For waves propagating in a linear, homogeneous and source-free medium, Eqs. (2.34a, 2.34b) are valid. When the two equations are combined, the determinant of the coefficient matrix should be zero, namely

$$\left| \overline{\overline{\kappa}}'_k - \left(\overline{\overline{\chi}}'_k - iu\overline{\overline{\sigma}}_y \right) \overline{\overline{\nu}}'^{-1}_k \left(\overline{\overline{\gamma}}'_k + iu\overline{\overline{\sigma}}_y \right) \right| = 0, \quad (2.35)$$

where

$$\overline{\overline{\kappa}}'_k = \begin{pmatrix} \kappa_{11} & \kappa_{12} \\ \kappa_{21} & \kappa_{22} \end{pmatrix}, \quad (2.36a)$$

$$\overline{\overline{\chi}}'_k = \begin{pmatrix} \chi_{11} & \chi_{12} \\ \chi_{21} & \chi_{22} \end{pmatrix}, \quad (2.36b)$$

$$\overline{\overline{\gamma}}'_k = \begin{pmatrix} \gamma_{11} & \gamma_{12} \\ \gamma_{21} & \gamma_{22} \end{pmatrix}, \quad (2.36c)$$

$$\overline{\overline{\nu}}'_k = \begin{pmatrix} \nu_{11} & \nu_{12} \\ \nu_{21} & \nu_{22} \end{pmatrix}, \quad (2.36d)$$

$$\overline{\overline{\sigma}}_y = \begin{pmatrix} 0 & -i \\ i & 0 \end{pmatrix}. \quad (2.36e)$$

By comparing Eq. (2.35) with Eq. (2.13), we find that the dispersion relation in the *kDB* coordinate system is simpler since the coefficients are 2×2 matrices. The benefit of using the *kDB* coordinate system will be more clear for the calculation of wave propagation in anisotropic media or even more complicated media.

2.4 Negative Refraction

Refraction is a universal phenomenon in nature. For isotropic electromagnetic materials, the refraction of a wave from one medium with refractive index n_1 to another medium with refractive index n_2 obeys Snell's law,

$$n_1 \sin\theta_1 = n_2 \sin\theta_2, \quad (2.37)$$

where θ_1 and θ_2 are the incident angle and refraction angle, respectively. For most dielectric media in nature, the refractive indices defined as $n = \sqrt{\varepsilon\mu}$ are positive. Thus, the refraction angle always has the same sign with the incident angle, namely the refraction is positive.

Contrary to the positive refraction, the refraction angle for negative refraction has a different sign from that of the incident angle. This requires that one of the two media have a negative refractive index. We will demonstrate in the next section that a medium with a negative permittivity and a negative permeability has a negative refractive index. Thus, searching for materials with both negative permittivity and negative permeability is required to achieve negative refraction.

2.5 Metamaterials

Many kinds of constitutive parameters cannot be realized in natural materials. However, the artificial metamaterials that people have proposed in recent years provide a feasible way to realize the unusual parameters [5].

2.5.1 Metamaterials with $\varepsilon < 0$

In the visible and near-ultraviolet regimes, the plasmon is a well-established collective excitation in metals and acts as the microscopic explanation of the negative permittivity of metals. The permittivity of metals can be described by a Drude model,

$$\varepsilon(\omega) = 1 - \frac{\omega_p^2}{\omega(\omega + i\gamma)}, \tag{2.38}$$

where ω_p is the plasma frequency and γ is a damping term representing the dissipation of the plasmon energy. For most metals the plasma frequency is typically in the ultraviolet region of the spectrum and γ is relatively small compared with ω_p. For instance, the parameters for aluminum are $\omega_p = 15$ eV and $\gamma = 0.1$ eV. One significant point is that the real part of the permittivity is negative below the plasma frequency, at least down to frequencies comparable to γ. However, at lower frequencies, as the dissipation parameter γ becomes comparable and even larger than the frequency ω, the permittivity would have a very large magnitude, which implies that all waves would decay within extremely short distances in the medium.

In order to depress the plasma frequency into the far infrared or even GHz band, Pendry et al. have proposed a strategy to dilute the average concentration of electrons by using periodic wire structures [6]. Consider a building block comprising very thin metallic wires of radius r placed periodically in a square lattice with the lattice constant a ($a \gg r$), as shown in Fig. 2.2. The external

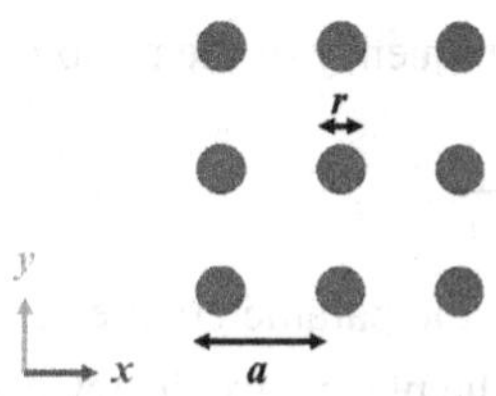

Figure 2.2 A periodic metallic wire array.

electric field is applied parallel to the wires along the z axis, and the electromagnetic response can be analyzed in the quasi-static limit if the wavelength is much larger than the lattice constant with $\lambda \gg a$. The density of the active electrons is given by the fraction of space occupied by the wire,

$$n_{\text{eff}} = \frac{\pi r^2}{a^2} n, \tag{2.39}$$

where n is the actual density of electrons in the wires. Before we move to the calculation of plasma frequency, we must first look at another effect that is important in the diluted wire array: the action of restoring force on the electrons would work against self-inductance of the wire structure. Suppose a current I_z flows in the wire and can generate a magnetic field circling the wire with

$$\overline{H}(\rho) = \hat{\phi} \frac{I_z}{2\pi\rho}, \tag{2.40}$$

where ρ is the distance from the wire center, $I_z = \pi r^2 nev$ is the current and v is the mean electron velocity. The magnetic field can also be written in terms of a vector potential as $\overline{H} = \mu_0^{-1} \nabla \times \overline{A}$. Then we can write the vector potential for a single wire as

$$\overline{A}(\rho) = \hat{z} \frac{\mu_0 I_z}{2\pi} \ln\left(\frac{a}{\rho}\right), \tag{2.41}$$

where the factor $\ln(a/\rho)$ is determined by considering that the vector potential should be zero at the boundary of each unit cell due to the array's symmetry. Since $a \gg r$ and the wire is supposed to be a perfect conductor, most of the electrons flow on the surface of the wire. Hence we can get the momentum per unit length of the wire as

$$\pi r^2 neA(r) = \pi r^2 ne \frac{\mu_0 I_z}{2\pi} \ln\left(\frac{a}{r}\right) = m_{\text{eff}} \pi r^2 nv, \tag{2.42}$$

where m_{eff} is the effective mass of the electrons given by

$$m_{\text{eff}} = \frac{\mu_0 \pi r^2 ne^2}{2\pi} \ln\left(\frac{a}{r}\right). \tag{2.43}$$

Then we obtain the plasma frequency of the periodic wire structure

$$\omega_p^2 = \frac{n_{\text{eff}} e^2}{\varepsilon_0 m_{\text{eff}}} = \frac{2\pi c^2}{a^2 \ln(a/r)}. \tag{2.44}$$

For the case of aluminum, if the parameters are set to $r = 1\ \mu\text{m}$, $a = 5\ \mu\text{m}$ and $n = 1.806 \times 10^{29}\ \text{m}^{-3}$, the plasma frequency can be reduced to 8.2 GHz.

Intriguingly, from Eq. (2.44) we see that the effective plasma frequency is not only related to the electron density in the wire but also depends on the wire radius and the lattice spacing. This means that the electromagnetic response can also be analyzed via the effective circuit model [7], assuming that the voltage generated per unit length along the z direction is U_z. According to the circuit theory, it can be written as $U_z = I_z(-i\omega L)$, where L is the total inductance per unit length and I_z is the current in a wire. On the other hand, the voltage is related to the average electric field, and therefore we can get the equation $E_z = I_z(-i\omega L)$. The polarization can then be written as

$$P_z = \frac{J_z}{-i\omega} = \frac{I_z}{-i\omega a^2} = -\frac{E_z}{\omega^2 a^2 L}. \tag{2.45}$$

Next we need to find the wire inductance, L. To do that, one needs to estimate the magnetic field in the wire medium. Due to the symmetry of the structure, the magnetic field should be zero in the middle between the wires. Hence the estimation of the magnetic flux can be written as

$$H_\phi(\rho) = \frac{I_z}{2\pi}\left(\frac{1}{\rho} - \frac{1}{a-\rho}\right), \tag{2.46}$$

where the two terms represent the quasi-static fields generated by the neighboring wires, respectively. Subsequently, the magnetic flux per unit length is obtained as

$$\psi = \mu_0 \int_r^{a/2} H_\phi(\rho)\, d\rho = \frac{\mu_0 I_z}{2\pi} \ln\left[\frac{a^2}{4r(a-r)}\right]. \tag{2.47}$$

Then the inductance can be calculated by

$$L = \frac{\psi}{I_z} = \frac{\mu_0}{2\pi} \ln\left[\frac{a^2}{4r(a-r)}\right], \tag{2.48}$$

and the effective permittivity is

$$\begin{aligned} \varepsilon_z(\omega) &= \frac{D_z}{E_z} = \frac{\varepsilon_0 E_z + P_z}{E_z} \\ &= \varepsilon_0 \left[1 - 2\pi\left(k_0^2 a^2 \ln\frac{a^2}{4r(a-r)}\right)^{-1}\right]. \end{aligned} \tag{2.49}$$

The above analysis assumes that only the inductance exists in the medium. For other periodic structures, such as the periodic cut-wire arrays, the circuit model would be more complicated, and the capacitance must be considered as well. Moreover, if the metal is lossy, one should also estimate the influence of the resistance. For this case the general relation between electric field and current becomes

$$E_z = I_z \left(-i\omega L + \sigma \pi r^2 + \frac{1}{-i\omega C} \right) \tag{2.50}$$

and the effective permittivity changes to

$$\varepsilon_z(\omega) = \frac{D_z}{E_z} = \varepsilon_0 \left[1 - \frac{\omega_p^2}{\omega^2 - 1/(LC) + i\omega\gamma} \right]. \tag{2.51}$$

Here ω_p is the reduced plasma frequency, $\omega_0 = 1/\sqrt{LC}$ is the resonant frequency and γ is the damping term.

It is worth noting that here we only discuss the simplest case of metallic wire arrays, where analytical solutions can be found. For other metamaterials with irregular shapes in periodic forms or random forms, numerical simulations in combination with retrieval algorithms could be an efficient approach to get the effective parameters when the sizes of the unit cells are much smaller than the operating wavelength.

2.5.2 Metamaterials with $\mu < 0$

The realization approach for negative permeability was first proposed by Pendry et al. in 1999 [8]. A wide range of effective permeabilities can be achieved by designing periodic microstructures from nonmagnetic thin sheets of metals. For instance, arrays of metallic cylinders can give us the magnetic properties of $\mu < 1$, while the split-ring resonators could provide the negative permeabilities.

In this section, we take the symmetric split-ring resonator (SRR) as an example in which the magnetic response can be efficiently tuned without bian-isotropic effect [4, 9]. The structure is shown in Fig. 2.3, where the ring has

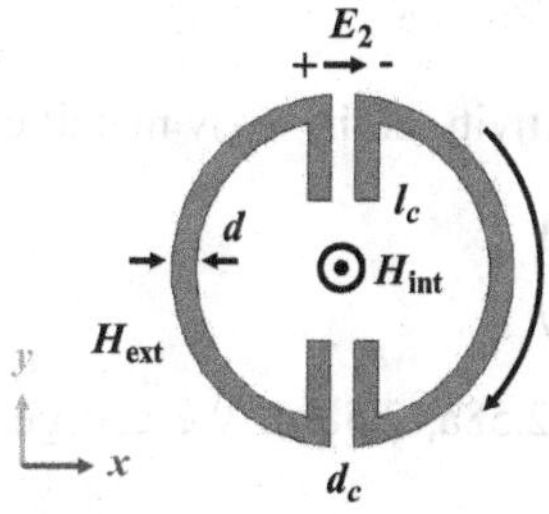

Figure 2.3 A split-ring resonator (SRR).

two gaps with gap length l_c and width d_c. The fractional area occupied by the split rings is f. Assume the magnetic field H_0 is applied in the z direction. Then a surface current, $J_\phi = H_{\text{ext}} - H_{\text{int}}$, is induced, where H_{ext} and H_{int} are the external and internal magnetic fields, respectively. The macroscopic average magnetic field is defined by averaging the field along each of the three axes of the unit cell. For our two-dimensional case, the average magnetic field H_{ave} equals the external field H_{ext}. Noting that

$$fH_{\text{int}} + (1 - f)\, H_{\text{ext}} = H_0, \tag{2.52}$$

then we get the magnetic fields

$$H_{\text{ext}} = H_0 + fJ_\phi, \tag{2.53}$$

$$H_{\text{int}} = H_0 + (1 - f)\, J_\phi, \tag{2.54}$$

and the effective permeability is

$$\mu_{\text{eff}} = \frac{B_{\text{ave}}}{H_{\text{ave}}} = \frac{\mu_0 H_0}{H_{\text{ext}}} = \mu_0 \left(1 - \frac{fJ_\phi}{H_{\text{int}} + J_\phi}\right). \tag{2.55}$$

Next we derive the expression of the internal magnetic field H_{int}. Over the closed loop of the split-ring resonator, the line integral of the electric field is the so-called electromotive force (EMF), and it is identical to magnetic flux within the enclosed area

$$\text{EMF} = \int_c d\bar{l} \cdot \bar{E} = i\omega \iint d\bar{S} \cdot \bar{B}. \tag{2.56}$$

Assuming that the electric fields in the metal and gap are E_1 and E_2, respectively, we find

$$-\left(2\pi R E_1 + 2d_c E_2\right) = i\omega\mu_0 \pi R^2 H_{\text{int}}, \tag{2.57}$$

where E_1 and E_2 can be derived to yield

$$J_\phi = d\frac{d}{dt}D_{1\phi} = -i\omega d\tilde{\varepsilon} E_1, \tag{2.58a}$$

$$J_\phi = l_c\frac{d}{dt}D_{2\phi} = -i\omega l_c \varepsilon_0 E_2. \tag{2.58b}$$

Here $\tilde{\varepsilon}$ is the complex permittivity of the lossy metallic wire. Let

$$\tilde{\varepsilon} = \varepsilon_0 \left(1 - \frac{\omega_p^2}{\omega\left(\omega + i\gamma\right)}\right) \tag{2.59}$$

and substitute it into Eqs. (2.58a, 2.58b). We can get the expression of the impedance

$$i\omega\mu_0 \pi R^2 H_{\text{int}} = -ZJ_\phi, \tag{2.60a}$$

$$Z = \left(-i\omega \frac{2\pi R}{d\varepsilon_0 \omega_p^2} + \gamma \frac{2\pi R}{d\varepsilon_0 \omega_p^2} + i \frac{2d_c}{\omega l_c \varepsilon_0} \right). \tag{2.60b}$$

Moreover, the impedance can be rewritten as a series of inductance, resistance and capacitance by

$$Z = -i\omega L_i + \gamma L_i + \frac{i}{\omega C}, \tag{2.61}$$

with $L_i = 2\pi R / \left(d\varepsilon_0 \omega_p^2 \right)$ and $C = l_c \varepsilon_0 / (2d_c)$. The internal and external magnetic fields are then obtained

$$H_{\text{int}} = \frac{iZJ_\phi}{\omega L_g}, \tag{2.62a}$$

$$H_{\text{ext}} = \left(1 + \frac{iZ}{\omega L_g} \right) J_\phi. \tag{2.62b}$$

We then obtain from Eq. (2.55) the effective permeability

$$\mu_{\text{eff}} = \mu_0 \left[1 - \frac{f}{1 + iZ / (\omega L_g)} \right] = \mu_0 \left[1 - \frac{f\omega^2 L_g / (L_g + L_i)}{\omega^2 - \omega_p^2 + i\omega\Gamma} \right], \tag{2.63}$$

where $\Gamma = \gamma L_i / (L_g + L_i)$ and $\omega_0^2 = 1 / \left[(L_g + L_i) C \right]$. In the case of a perfect conductor where $\Gamma = 0$, we have

$$\frac{\mu_{\text{eff}}}{\mu_0} = \frac{\omega^2 \left[1 - fL_g / (L_g + L_i) \right] - \omega_0^2}{\omega^2 - \omega_0^2}. \tag{2.64}$$

Hence the effective permeability is negative within the frequency range

$$\omega_0^2 \leq \omega^2 \leq \frac{\omega_0^2}{1 - fL_g / (L_g + L_i)}. \tag{2.65}$$

2.5.3 Metamaterials with Both $\varepsilon < 0$ and $\mu < 0$

In previous sections we introduced the theoretical basis of metamaterials with either negative permittivity or negative permeability. A diluted metallic wire array could efficiently reduce the plasma frequency and hence give us small negative permittivity in low-frequency regions. Negative permeability is achieved from the magnetic resonant mode of the split-ring resonators. However, only an evanescent wave can exist in a medium with only negative permittivity or negative permeability, and therefore it shows a forbidden band. One can achieve a passband with negative refraction by combining the wire media and the split-ring resonators, assuming that the plasma frequency of a wire media is reduced to ω_p with the effective dielectric function of

$$\varepsilon_{\text{eff}} = \varepsilon_0 \left(1 - \frac{\omega_p^2}{\omega^2} \right). \tag{2.66}$$

Similarly, the effective permeability of a split-ring resonator is

$$\mu_{\text{eff}} = \mu_0 \left(1 - \frac{F\omega^2}{\omega^2 - \omega_0^2}\right) \tag{2.67}$$

in which we neglect the loss. The dispersion relation of the combined media can be derived as

$$k^2 = \omega^2 \varepsilon_{\text{eff}} \mu_{\text{eff}} = \frac{\left(\omega^2 - \omega_p^2\right)\left(\omega^2 - F\omega^2 - \omega_0^2\right)}{c^2\left(\omega^2 - \omega_0^2\right)}. \tag{2.68}$$

This equation indicates that the range of the passband (real k) extends from ω_0 to ω_{np}, where $\omega_{np} = \omega_0/\sqrt{1-F}$. In this passband the direction of the group velocity is opposite of that of the phase velocity, and negative refraction occurs.

3 Negative Refraction in Isotropic Metamaterials

3.1 Introduction

Isotropic metamaterials have the simplest constitutive parameters, where each of the two constitutive parameter tensors has only one element, respectively. In this section we will introduce the basic theoretical concepts and design principles of negative refraction in isotropic metamaterials.

3.2 Waves in Isotropic Media

The source-free Maxwell equations in isotropic media in the *xyz* coordinate system are

$$\overline{k} \times \overline{E} = \omega \overline{B}, \tag{3.1a}$$

$$\overline{k} \times \overline{H} = -\omega \overline{D}, \tag{3.1b}$$

$$\overline{k} \cdot \overline{D} = 0, \tag{3.1c}$$

$$\overline{k} \cdot \overline{B} = 0, \tag{3.1d}$$

and the constitutive relations are

$$\overline{D} = \varepsilon \overline{E}, \tag{3.2a}$$

$$\overline{B} = \mu \overline{H}, \tag{3.2b}$$

in the $\overline{EH}$ representation, where ε is the permittivity and μ is the permeability. In the following we will derive the dispersion relations of waves in isotropic media in the *kDB* coordinate system.

First, we need to transform the constitutive relations from the $\overline{EH}$ representation in the *xyz* coordinate system to the $\overline{DB}$ representation in the *xyz* coordinate system. According to Eqs. (2.25a, 2.25b, 2.26a, 2.26b, 2.26c, 2.26d, 2.27a,

2.27b, 2.27c, 2.27d), the constitutive relations in the $\overline{DB}$ representation in the *xyz* coordinate system are

$$\overline{E} = \kappa\overline{D}, \tag{3.3a}$$
$$\overline{H} = \nu\overline{B}, \tag{3.3b}$$

where

$$\kappa = 1/\varepsilon, \tag{3.4a}$$
$$\nu = 1/\mu, \tag{3.4b}$$

and

$$\varepsilon = 1/\kappa, \tag{3.5a}$$
$$\mu = 1/\nu. \tag{3.5b}$$

Second, we need to transform the Maxwell equations and constitutive relations from the *xyz* coordinate system to the *kDB* coordinate system. From Eqs. (2.32a, 2.32b, 2.32c, 2.32d), the Maxwell equations in isotropic media in the *kDB* coordinate system are

$$\overline{k}_k \times \overline{E}_k = \omega\overline{B}_k, \tag{3.6a}$$
$$\overline{k}_k \times \overline{H}_k = -\omega\overline{D}_k, \tag{3.6b}$$
$$\overline{k}_k \cdot \overline{D}_k = 0, \tag{3.6c}$$
$$\overline{k}_k \cdot \overline{B}_k = 0, \tag{3.6d}$$

and the constitutive relations are

$$\overline{E}_k = \kappa_k\overline{D}_k, \tag{3.7a}$$
$$\overline{H}_k = \nu_k\overline{B}_k, \tag{3.7b}$$

where

$$\kappa_k = \kappa, \tag{3.8a}$$
$$\nu_k = \nu, \tag{3.8b}$$

according to Eqs. (2.28a, 2.28b, 2.29a, 2.29b, 2.29c, 2.29d).

Third, we need to derive the dispersion relations from the coefficient matrix. According to Eqs. (2.35, 2.36a, 2.36b, 2.36c, 2.36d, 3.8a, 3.8b), the determinant of the coefficient matrix is

$$\left|\overline{\overline{\kappa}}'_k - u^2\overline{\overline{\nu}}'^{-1}_k\right| = 0, \tag{3.9}$$

where

$$\overline{\overline{\kappa}}_k^{\prime} = \begin{pmatrix} \kappa_k & 0 \\ 0 & \kappa_k \end{pmatrix} = \begin{pmatrix} \kappa & 0 \\ 0 & \kappa \end{pmatrix}, \tag{3.10a}$$

$$\overline{\overline{\nu}}_k^{\prime} = \begin{pmatrix} \nu_k & 0 \\ 0 & \nu_k \end{pmatrix} = \begin{pmatrix} \nu & 0 \\ 0 & \nu \end{pmatrix}, \tag{3.10b}$$

and $u = \omega/k$. Thus the dispersion relation is

$$u = \sqrt{\kappa\nu}. \tag{3.11}$$

According to Eqs. (2.34a, 2.34b), the degenerate dispersion relation corresponds to two types of characteristic waves with $D_1 \neq 0$, $D_2 = 0$ and $D_1 = 0$, $D_2 \neq 0$, respectively. We define the waves with $D_1 \neq 0$ and $D_2 = 0$ as the *type-I waves*, and the waves with $D_1 = 0$ and $D_2 \neq 0$ as the *type-II waves*. It is worth noting that in this section we let $\kappa\nu > 0$ for both the type-I and type-II waves in order to study the propagating waves.

Fourth, we study the type-I and type-II waves in the *kDB* coordinate system, respectively. For a type-I wave, the wave vector is

$$\overline{k}_k^{\mathrm{I}} = \hat{e}_3 k^{\mathrm{I}} = \hat{e}_3 \frac{\omega}{\sqrt{\kappa\nu}}, \tag{3.12}$$

which is always positive. Since $D_1 \neq 0$ and $D_2 = 0$, the electric flux density is

$$\overline{D}_k^{\mathrm{I}} = \hat{e}_1 D^{\mathrm{I}}. \tag{3.13a}$$

According to Eq. (2.34a) or Eq. (2.34b), the magnetic flux density is

$$\overline{B}_k^{\mathrm{I}} = \begin{cases} \hat{e}_2 \sqrt{\kappa/\nu} D^{\mathrm{I}}, & \kappa > 0 \text{ and } \nu > 0, \\ -\hat{e}_2 \sqrt{\kappa/\nu} D^{\mathrm{I}}, & \kappa < 0 \text{ and } \nu < 0. \end{cases} \tag{3.13b}$$

According to Eq. (3.7a), the electric field is

$$\overline{E}_k^{\mathrm{I}} = \hat{e}_1 \kappa D^{\mathrm{I}}. \tag{3.13c}$$

Then according to Eq. (3.7b), the magnetic field is

$$\overline{H}_k^{\mathrm{I}} = \hat{e}_2 \sqrt{\kappa\nu} D^{\mathrm{I}}. \tag{3.13d}$$

Similarly, for a type-II wave, the wave vector is

$$\overline{k}_k^{\mathrm{II}} = \hat{e}_3 k^{\mathrm{II}} = \hat{e}_3 \frac{\omega}{\sqrt{\kappa\nu}}, \tag{3.14}$$

and the field components are

$$\overline{D}_k^{\mathrm{II}} = \hat{e}_2 D^{\mathrm{II}}, \tag{3.15a}$$

$$\overline{B}_k^{\mathrm{II}} = \begin{cases} -\hat{e}_1\sqrt{\kappa/\nu}D^{\mathrm{II}}, & \kappa > 0 \text{ and } \nu > 0, \\ \hat{e}_1\sqrt{\kappa/\nu}D^{\mathrm{II}}, & \kappa < 0 \text{ and } \nu < 0, \end{cases} \tag{3.15b}$$

$$\overline{E}_k^{\mathrm{II}} = \hat{e}_2\kappa D^{\mathrm{II}}, \tag{3.15c}$$

$$\overline{H}_k^{\mathrm{II}} = -\hat{e}_1\sqrt{\kappa\nu}D^{\mathrm{II}}. \tag{3.15d}$$

The Poynting's vectors in the *kDB* coordinate system for both the type-I and type-II waves can be calculated by

$$\langle \overline{S}_k \rangle = \frac{1}{2}\mathrm{Re}\left(\overline{E}_k \times \overline{H}_k^*\right). \tag{3.16}$$

From Eqs. (3.13c, 3.13d), for a type-I wave, we get

$$\left\langle \overline{S}_k^{\mathrm{I}} \right\rangle = \begin{cases} \hat{e}_3\frac{\kappa\nu}{2}\sqrt{\frac{\kappa}{\nu}}|D^{\mathrm{I}}|^2, & \kappa > 0 \text{ and } \nu > 0, \\ -\hat{e}_3\frac{\kappa\nu}{2}\sqrt{\frac{\kappa}{\nu}}|D^{\mathrm{I}}|^2, & \kappa < 0 \text{ and } \nu < 0. \end{cases} \tag{3.17}$$

Similarly, from Eqs. (3.15c, 3.15d), for a type-II wave, we get

$$\left\langle \overline{S}_k^{\mathrm{II}} \right\rangle = \begin{cases} \hat{e}_3\frac{\kappa\nu}{2}\sqrt{\frac{\kappa}{\nu}}|D^{\mathrm{II}}|^2, & \kappa > 0 \text{ and } \nu > 0, \\ -\hat{e}_3\frac{\kappa\nu}{2}\sqrt{\frac{\kappa}{\nu}}|D^{\mathrm{II}}|^2, & \kappa < 0 \text{ and } \nu < 0. \end{cases} \tag{3.18}$$

Thus, for any type of waves, there are two sets of field components, which correspond to $\kappa > 0$, $\nu > 0$ and $\kappa < 0$, $\nu < 0$, respectively. In the following we define the media with $\kappa > 0$ and $\nu > 0$ as the *conventional materials*, and the media with $\kappa < 0$ and $\nu < 0$ as the *metamaterials*. The type-I and type-II waves constitute a set of eigenstates in isotropic media, and any waves can be expanded using this set of eigenstates.

Finally, we transform the field components from the *kDB* coordinate system back to the *xyz* coordinate system. The coordinate systems are shown in Fig. 2.1. According to Eqs. (2.22a, 2.22b, 2.22c, 2.22d), for the type-I waves, we have

$$\overline{D}^{\mathrm{I}} = \hat{x}\sin\phi D^{\mathrm{I}} - \hat{y}\cos\phi D^{\mathrm{I}}, \tag{3.19a}$$

$$\overline{E}^{\mathrm{I}} = \hat{x}\sin\phi\kappa D^{\mathrm{I}} - \hat{y}\cos\phi\kappa D^{\mathrm{I}}, \tag{3.19b}$$

$$\overline{B}^{\mathrm{I}} = \hat{x}\cos\theta\cos\phi\sqrt{\kappa/\nu}D^{\mathrm{I}} + \hat{y}\cos\theta\sin\phi\sqrt{\kappa/\nu}D^{\mathrm{I}} - \hat{z}\sin\theta\sqrt{\kappa/\nu}D^{\mathrm{I}}, \tag{3.19c}$$

$$\overline{H}^{\mathrm{I}} = \hat{x}\cos\theta\cos\phi\sqrt{\kappa\nu}D^{\mathrm{I}} + \hat{y}\cos\theta\sin\phi\sqrt{\kappa\nu}D^{\mathrm{I}} - \hat{z}\sin\theta\sqrt{\kappa\nu}D^{\mathrm{I}}, \tag{3.19d}$$

in the conventional materials, and

$$\overline{D}^{\mathrm{I}} = \hat{x}\sin\phi D^{\mathrm{I}} - \hat{y}\cos\phi D^{\mathrm{I}}, \tag{3.20a}$$

$$\overline{E}^{\mathrm{I}} = \hat{x}\sin\phi\kappa D^{\mathrm{I}} - \hat{y}\cos\phi\kappa D^{\mathrm{I}}, \tag{3.20b}$$

$$\overline{B}^{\mathrm{I}} = -\hat{x}\cos\theta\cos\phi\sqrt{\kappa/\nu}D^{\mathrm{I}} - \hat{y}\cos\theta\sin\phi\sqrt{\kappa/\nu}D^{\mathrm{I}} + \hat{z}\sin\theta\sqrt{\kappa/\nu}D^{\mathrm{I}}, \tag{3.20c}$$

$$\overline{H}^{\mathrm{I}} = \hat{x}\cos\theta\cos\phi\sqrt{\kappa\nu}D^{\mathrm{I}} + \hat{y}\cos\theta\sin\phi\sqrt{\kappa\nu}D^{\mathrm{I}} - \hat{z}\sin\theta\sqrt{\kappa\nu}D^{\mathrm{I}}, \tag{3.20d}$$

in the metamaterials, respectively. Similarly, for the type-II waves, we have

$$\overline{D}^{\mathrm{II}} = \hat{x}\cos\theta\cos\phi D^{\mathrm{II}} + \hat{y}\cos\theta\sin\phi D^{\mathrm{II}} - \hat{z}\sin\theta D^{\mathrm{II}}, \tag{3.21a}$$

$$\overline{E}^{\mathrm{II}} = \hat{x}\cos\theta\cos\phi\kappa D^{\mathrm{II}} + \hat{y}\cos\theta\sin\phi\kappa D^{\mathrm{II}} - \hat{z}\sin\theta\kappa D^{\mathrm{II}}, \tag{3.21b}$$

$$\overline{B}^{\mathrm{II}} = -\hat{x}\sin\phi\sqrt{\kappa/\nu}D^{\mathrm{II}} + \hat{y}\cos\phi\sqrt{\kappa/\nu}D^{\mathrm{II}}, \tag{3.21c}$$

$$\overline{H}^{\mathrm{II}} = -\hat{x}\sin\phi\sqrt{\kappa\nu}D^{\mathrm{II}} + \hat{y}\cos\phi\sqrt{\kappa\nu}D^{\mathrm{II}}, \tag{3.21d}$$

in the conventional materials, and

$$\overline{D}^{\mathrm{II}} = \hat{x}\cos\theta\cos\phi D^{\mathrm{II}} + \hat{y}\cos\theta\sin\phi D^{\mathrm{II}} - \hat{z}\sin\theta D^{\mathrm{II}}, \tag{3.22a}$$

$$\overline{E}^{\mathrm{II}} = \hat{x}\cos\theta\cos\phi\kappa D^{\mathrm{II}} + \hat{y}\cos\theta\sin\phi\kappa D^{\mathrm{II}} - \hat{z}\sin\theta\kappa D^{\mathrm{II}}, \tag{3.22b}$$

$$\overline{B}^{\mathrm{II}} = \hat{x}\sin\phi\sqrt{\kappa/\nu}D^{\mathrm{II}} - \hat{y}\cos\phi\sqrt{\kappa/\nu}D^{\mathrm{II}}, \tag{3.22c}$$

$$\overline{H}^{\mathrm{II}} = -\hat{x}\sin\phi\sqrt{\kappa\nu}D^{\mathrm{II}} + \hat{y}\cos\phi\sqrt{\kappa\nu}D^{\mathrm{II}}, \tag{3.22d}$$

in the metamaterials, respectively. Note that the above field components are expressed in the $\overline{DB}$ representation.

When transformed from the $\overline{DB}$ representation to the $\overline{EH}$ representation, the dispersion relations are

$$k = \omega\sqrt{\varepsilon\mu}, \tag{3.23}$$

according to Eqs. (3.4a, 3.4b, 3.11). We get two cases in total. For case 1, the conventional materials with $\kappa > 0$ and $\nu > 0$ require that $\varepsilon > 0$ and $\mu > 0$. For case 2, the metamaterials with $\kappa < 0$ and $\nu < 0$ require that $\varepsilon < 0$ and $\mu < 0$. For a type-I wave, the wave number is

$$k^{\mathrm{I}} = \omega\sqrt{\varepsilon\mu}, \tag{3.24}$$

which is always positive. The field components are omitted here for brevity. Readers can easily derive them by themselves. Considering $\overline{k} = \hat{e}_3 k = \hat{x}k\sin\theta\cos\phi + \hat{y}k\sin\theta\sin\phi + \hat{z}k\cos\theta$ and $\overline{r} = \hat{x}x + \hat{y}y + \hat{z}z$, the field components contain a phase factor, where

$$D^{\mathrm{I}} = D_0^{\mathrm{I}}e^{i\overline{k}^{\mathrm{I}}\cdot\overline{r}} = D_0^{\mathrm{I}}e^{ik^{\mathrm{I}}(x\sin\theta\cos\phi + y\sin\theta\sin\phi + z\cos\theta)}. \tag{3.25}$$

Similarly, for a type-II wave, the wave number is

$$k^{\mathrm{II}} = \omega\sqrt{\varepsilon\mu}, \tag{3.26}$$

Table 3.1 The main properties of type-I waves in isotropic media.

type-I wave	**direction of $\overline{k}^{I}$**	**direction of $\left\langle \overline{S}^{I} \right\rangle$**
case 1: $\varepsilon > 0$ and $\mu > 0$	$\hat{e}_3$	$\hat{e}_3$
case 2: $\varepsilon < 0$ and $\mu < 0$	$\hat{e}_3$	$-\hat{e}_3$

Table 3.2 The main properties of type-II waves in isotropic media.

type-II wave	**direction of $\overline{k}^{II}$**	**direction of $\left\langle \overline{S}^{II} \right\rangle$**
case 1: $\varepsilon > 0$ and $\mu > 0$	$\hat{e}_3$	$\hat{e}_3$
case 2: $\varepsilon < 0$ and $\mu < 0$	$\hat{e}_3$	$-\hat{e}_3$

which is also always positive. The field components are also omitted. The field components contain a phase factor, where

$$D^{II} = D_0^{II} e^{i\overline{k}^{II} \cdot \overline{r}} = D_0^{II} e^{ik^{II}(x \sin\theta \cos\phi + y \sin\theta \sin\phi + z \cos\theta)}. \tag{3.27}$$

The Poynting's vectors in the *xyz* coordinate system can be calculated as

$$\langle \overline{S} \rangle = \frac{1}{2} \text{Re} \left(\overline{E} \times \overline{H}^* \right). \tag{3.28}$$

For the type-I waves, we get

$$\left\langle \overline{S}^{I} \right\rangle = \begin{cases} \hat{e}_3 \frac{\eta}{2\varepsilon\mu} D_0^{I2}, & \varepsilon > 0 \text{ and } \mu > 0, \\ -\hat{e}_3 \frac{\eta}{2\varepsilon\mu} D_0^{I2}, & \varepsilon < 0 \text{ and } \mu < 0, \end{cases} \tag{3.29}$$

where $\hat{e}_3 = \hat{x} \sin\theta \cos\phi + \hat{y} \sin\theta \sin\phi + \hat{z} \cos\theta$. Similarly, for the type-II waves, we get

$$\left\langle \overline{S}^{II} \right\rangle = \begin{cases} \hat{e}_3 \frac{\eta}{2\varepsilon\mu} D_0^{II2}, & \varepsilon > 0 \text{ and } \mu > 0, \\ -\hat{e}_3 \frac{\eta}{2\varepsilon\mu} D_0^{II2}, & \varepsilon < 0 \text{ and } \mu < 0. \end{cases} \tag{3.30}$$

In light of the directions of the wave vectors and Poynting's vectors, the main properties of type-I and type-II waves are summarized in Tabs. 3.1 and 3.2.

At the end of this section we would like to conclude the study procedure of the wave types and wave propagations in the *kDB* coordinate system. Originally, the Maxwell equations were written in the *xyz* coordinate system, and the constitutive relations were expressed in the $\overline{EH}$ representation. In order to do the remaining calculations, the constitutive relations should be transformed from the $\overline{EH}$ representation in the *xyz* coordinate system to the $\overline{DB}$ representation in the *xyz* coordinate system. Then we can transform the Maxwell equations and

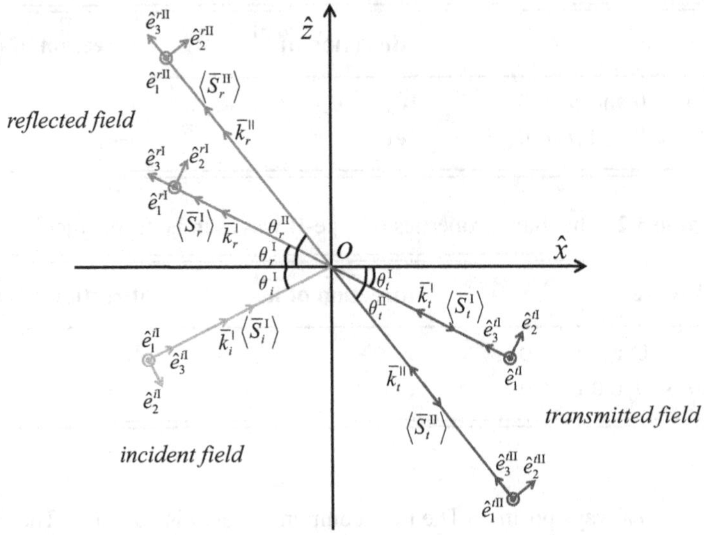

Figure 3.1 A schematic of the negative refraction at an interface between an isotropic medium in case 1 and an isotropic medium in case 2.

constitutive relations from the *xyz* coordinate system to the *kDB* coordinate system. In the *kDB* coordinate system, the dispersion relations and different types of waves can be obtained. And the field components and Poynting's vectors can be calculated to study the wave propagations. After we get these results, the field components for different types of waves can be transformed from the *kDB* coordinate system back to the *xyz* coordinate system, then from the $\overline{DB}$ representation back to the $\overline{EH}$ representation. At this point we can now study the wave types and wave propagations in the $\overline{EH}$ representation.

3.3 Negative Refraction in Isotropic Metamaterials

In this section we show that negative refraction can occur at an interface between two isotropic media. We let the incident plane be the *xz* plane and the boundary between the two media be the *yz* plane, as shown in Fig. 3.1. All the angles are assumed to have positive values.

As an example we only consider the incidence of a type-I wave and its refraction from an isotropic medium in case 1 with $\varepsilon > 0$ and $\mu > 0$ to an isotropic medium in case 2 with $\varepsilon < 0$ and $\mu < 0$. The field components and Poynting's vectors for both the type-I and type-II waves in both the media were derived at the end of the last section. For the incident field, $\theta = \pi/2 - \theta_i^{\mathrm{I}}$ and $\phi = 0$.

Then the incident field components are

$$\overline{D}_i^{\mathrm{I}} = -\hat{y} D_i^{\mathrm{I}} e^{ik_1^{\mathrm{I}}\left(x\cos\theta_i^{\mathrm{I}} + z\sin\theta_i^{\mathrm{I}}\right)}, \tag{3.31a}$$

$$\overline{E}_i^{\mathrm{I}} = -\hat{y}\frac{1}{\varepsilon_1} D_i^{\mathrm{I}} e^{ik_1^{\mathrm{I}}\left(x\cos\theta_i^{\mathrm{I}} + z\sin\theta_i^{\mathrm{I}}\right)}, \tag{3.31b}$$

$$\overline{B}_i^{\mathrm{I}} = \left(\hat{x}\sin\theta_i^{\mathrm{I}} - \hat{z}\cos\theta_i^{\mathrm{I}}\right)\eta_1 D_i^{\mathrm{I}} e^{ik_1^{\mathrm{I}}\left(x\cos\theta_i^{\mathrm{I}} + z\sin\theta_i^{\mathrm{I}}\right)}, \tag{3.31c}$$

$$\overline{H}_i^{\mathrm{I}} = \left(\hat{x}\sin\theta_i^{\mathrm{I}} - \hat{z}\cos\theta_i^{\mathrm{I}}\right)\frac{1}{\sqrt{\varepsilon_1\mu_1}} D_i^{\mathrm{I}} e^{ik_1^{\mathrm{I}}\left(x\cos\theta_i^{\mathrm{I}} + z\sin\theta_i^{\mathrm{I}}\right)}, \tag{3.31d}$$

and the Poynting's vector is

$$\left\langle\overline{S}_i^{\mathrm{I}}\right\rangle = \left(\hat{x}\cos\theta_i^{\mathrm{I}} + \hat{z}\sin\theta_i^{\mathrm{I}}\right)\frac{\eta_1}{2\varepsilon_1\mu_1} D_i^{\mathrm{I}2}, \tag{3.32}$$

where $\eta_1 = \sqrt{\mu_1/\varepsilon_1}$ is the impedance and $k_1^{\mathrm{I}} = \omega\sqrt{\varepsilon_1\mu_1}$ is positive. For the reflected fields, $\theta = \pi/2 - \theta_r^{\mathrm{I}}$ and $\phi = \pi$ for the type-I wave; and $\theta = \pi/2 - \theta_r^{\mathrm{II}}$ and $\phi = \pi$ for the type-II wave. Then the reflected field components for the type-I wave are

$$\overline{D}_r^{\mathrm{I}} = \hat{y} D_r^{\mathrm{I}} e^{ik_1^{\mathrm{I}}\left(-x\cos\theta_r^{\mathrm{I}} + z\sin\theta_r^{\mathrm{I}}\right)}, \tag{3.33a}$$

$$\overline{E}_r^{\mathrm{I}} = \hat{y}\frac{1}{\varepsilon_1} D_r^{\mathrm{I}} e^{ik_1^{\mathrm{I}}\left(-x\cos\theta_r^{\mathrm{I}} + z\sin\theta_r^{\mathrm{I}}\right)}, \tag{3.33b}$$

$$\overline{B}_r^{\mathrm{I}} = \left(-\hat{x}\sin\theta_r^{\mathrm{I}} - \hat{z}\cos\theta_r^{\mathrm{I}}\right)\eta_1 D_r^{\mathrm{I}} e^{ik_1^{\mathrm{I}}\left(-x\cos\theta_r^{\mathrm{I}} + z\sin\theta_r^{\mathrm{I}}\right)}, \tag{3.33c}$$

$$\overline{H}_r^{\mathrm{I}} = \left(-\hat{x}\sin\theta_r^{\mathrm{I}} - \hat{z}\cos\theta_r^{\mathrm{I}}\right)\frac{1}{\sqrt{\varepsilon_1\mu_1}} D_r^{\mathrm{I}} e^{ik_1^{\mathrm{I}}\left(-x\cos\theta_r^{\mathrm{I}} + z\sin\theta_r^{\mathrm{I}}\right)}, \tag{3.33d}$$

and the Poynting's vector is

$$\left\langle\overline{S}_r^{\mathrm{I}}\right\rangle = \left(-\hat{x}\cos\theta_r^{\mathrm{I}} + \hat{z}\sin\theta_r^{\mathrm{I}}\right)\frac{\eta_1}{2\varepsilon_1\mu_1} D_r^{\mathrm{I}2}. \tag{3.34}$$

Without loss of generality, here we assume that the type-II wave also exists for the reflected field. The field components are

$$\overline{D}_r^{\mathrm{II}} = \left(-\hat{x}\sin\theta_r^{\mathrm{II}} - \hat{z}\cos\theta_r^{\mathrm{II}}\right) D_r^{\mathrm{II}} e^{ik_1^{\mathrm{II}}\left(-x\cos\theta_r^{\mathrm{II}} + z\sin\theta_r^{\mathrm{II}}\right)}, \tag{3.35a}$$

$$\overline{E}_r^{\mathrm{II}} = \left(-\hat{x}\sin\theta_r^{\mathrm{II}} - \hat{z}\cos\theta_r^{\mathrm{II}}\right)\frac{1}{\varepsilon_1} D_r^{\mathrm{II}} e^{ik_1^{\mathrm{II}}\left(-x\cos\theta_r^{\mathrm{II}} + z\sin\theta_r^{\mathrm{II}}\right)}, \tag{3.35b}$$

$$\overline{B}_r^{\mathrm{II}} = -\hat{y}\eta_1 D_r^{\mathrm{II}} e^{ik_1^{\mathrm{II}}\left(-x\cos\theta_r^{\mathrm{II}} + z\sin\theta_r^{\mathrm{II}}\right)}, \tag{3.35c}$$

$$\overline{H}_r^{\mathrm{II}} = -\hat{y}\frac{1}{\sqrt{\varepsilon_1\mu_1}} D_r^{\mathrm{II}} e^{ik_1^{\mathrm{II}}\left(-x\cos\theta_r^{\mathrm{II}} + z\sin\theta_r^{\mathrm{II}}\right)}, \tag{3.35d}$$

and the Poynting's vector is

$$\left\langle\overline{S}_r^{\mathrm{II}}\right\rangle = \left(-\hat{x}\cos\theta_r^{\mathrm{II}} + \hat{z}\sin\theta_r^{\mathrm{II}}\right)\frac{\eta_1}{2\varepsilon_1\mu_1} D_r^{\mathrm{II}2}, \tag{3.36}$$

where $k_1^{\mathrm{II}} = \omega\sqrt{\varepsilon_1\mu_1}$ is positive. For the transmitted fields, $\theta = \pi/2 - \theta_t^{\mathrm{I}}$ and $\phi = \pi$ for the type-I wave; and $\theta = \pi/2 - \theta_t^{\mathrm{II}}$ and $\phi = \pi$ for the type-II wave by considering the directions of the energy flows. Then the transmitted field components for the type-I wave are

$$\overline{D}_t^{\mathrm{I}} = \hat{y} D_t^{\mathrm{I}} e^{ik_2^{\mathrm{I}}(-x\cos\theta_t^{\mathrm{I}} + z\sin\theta_t^{\mathrm{I}})}, \tag{3.37a}$$

$$\overline{E}_t^{\mathrm{I}} = \hat{y}\frac{1}{\varepsilon_2} D_t^{\mathrm{I}} e^{ik_2^{\mathrm{I}}(-x\cos\theta_t^{\mathrm{I}} + z\sin\theta_t^{\mathrm{I}})}, \tag{3.37b}$$

$$\overline{B}_t^{\mathrm{I}} = \left(\hat{x}\sin\theta_t^{\mathrm{I}} + \hat{z}\cos\theta_t^{\mathrm{I}}\right)\eta_2 D_t^{\mathrm{I}} e^{ik_2^{\mathrm{I}}(-x\cos\theta_t^{\mathrm{I}} + z\sin\theta_t^{\mathrm{I}})}, \tag{3.37c}$$

$$\overline{H}_t^{\mathrm{I}} = \left(-\hat{x}\sin\theta_t^{\mathrm{I}} - \hat{z}\cos\theta_t^{\mathrm{I}}\right)\frac{1}{\sqrt{\varepsilon_2\mu_2}} D_t^{\mathrm{I}} e^{ik_2^{\mathrm{I}}(-x\cos\theta_t^{\mathrm{I}} + z\sin\theta_t^{\mathrm{I}})}, \tag{3.37d}$$

and the Poynting's vector is

$$\left\langle \overline{S}_t^{\mathrm{I}} \right\rangle = -\left(-\hat{x}\cos\theta_t^{\mathrm{I}} + \hat{z}\sin\theta_t^{\mathrm{I}}\right)\frac{\eta_2}{2\varepsilon_2\mu_2} D_t^{\mathrm{I}2}, \tag{3.38}$$

where $\eta_2 = \sqrt{\mu_2/\varepsilon_2}$ is the impedance and $k_2^{\mathrm{I}} = \omega\sqrt{\varepsilon_2\mu_2}$ is positive. Similarly, the transmitted field components for the type-II wave are

$$\overline{D}_t^{\mathrm{II}} = \left(-\hat{x}\sin\theta_t^{\mathrm{II}} - \hat{z}\cos\theta_t^{\mathrm{II}}\right) D_t^{\mathrm{II}} e^{ik_2^{\mathrm{II}}(-x\cos\theta_t^{\mathrm{II}} + z\sin\theta_t^{\mathrm{II}})}, \tag{3.39a}$$

$$\overline{E}_t^{\mathrm{II}} = \left(-\hat{x}\sin\theta_t^{\mathrm{II}} - \hat{z}\cos\theta_t^{\mathrm{II}}\right)\frac{1}{\varepsilon_2} D_t^{\mathrm{II}} e^{ik_2^{\mathrm{II}}(-x\cos\theta_t^{\mathrm{II}} + z\sin\theta_t^{\mathrm{II}})}, \tag{3.39b}$$

$$\overline{B}_t^{\mathrm{II}} = \hat{y}\eta_2 D_t^{\mathrm{II}} e^{ik_2^{\mathrm{II}}(-x\cos\theta_t^{\mathrm{II}} + z\sin\theta_t^{\mathrm{II}})}, \tag{3.39c}$$

$$\overline{H}_t^{\mathrm{II}} = -\hat{y}\frac{1}{\sqrt{\varepsilon_2\mu_2}} D_t^{\mathrm{II}} e^{ik_2^{\mathrm{II}}(-x\cos\theta_t^{\mathrm{II}} + z\sin\theta_t^{\mathrm{II}})}, \tag{3.39d}$$

and the Poynting's vector is

$$\left\langle \overline{S}_t^{\mathrm{II}} \right\rangle = -\left(-\hat{x}\cos\theta_t^{\mathrm{II}} + \hat{z}\sin\theta_t^{\mathrm{II}}\right)\frac{\eta_2}{2\varepsilon_2\mu_2} D_t^{\mathrm{II}2}, \tag{3.40}$$

where $k_2^{\mathrm{II}} = \omega\sqrt{\varepsilon_2\mu_2}$ is positive. Note that the direction of the energy flow is always from the interface to infinity.

Considering the continuity conditions for the electric fields and magnetic fields, we have

$$-\frac{1}{\varepsilon_1} D_i^{\mathrm{I}} + \frac{1}{\varepsilon_1} D_r^{\mathrm{I}} = \frac{1}{\varepsilon_2} D_t^{\mathrm{I}}, \tag{3.41a}$$

$$\frac{\cos\theta_r^{\mathrm{II}}}{\varepsilon_1} D_r^{\mathrm{II}} = \frac{\cos\theta_t^{\mathrm{II}}}{\varepsilon_2} D_t^{\mathrm{II}}, \tag{3.41b}$$

$$\frac{1}{\sqrt{\varepsilon_1\mu_1}} D_r^{\mathrm{II}} = \frac{1}{\sqrt{\varepsilon_2\mu_2}} D_t^{\mathrm{II}}, \tag{3.41c}$$

$$\frac{\cos\theta_i^{\mathrm{I}}}{\sqrt{\varepsilon_1\mu_1}} D_i^{\mathrm{I}} + \frac{\cos\theta_r^{\mathrm{I}}}{\sqrt{\varepsilon_1\mu_1}} D_r^{\mathrm{I}} = \frac{\cos\theta_t^{\mathrm{I}}}{\sqrt{\varepsilon_2\mu_2}} D_t^{\mathrm{I}}, \tag{3.41d}$$

and

$$k_1^{\mathrm{I}} \sin\theta_i^{\mathrm{I}} = k_1^{\mathrm{I}} \sin\theta_r^{\mathrm{I}} = k_2^{\mathrm{I}} \sin\theta_t^{\mathrm{I}}, \tag{3.42a}$$

$$k_1^{\mathrm{II}} \sin\theta_r^{\mathrm{II}} = k_2^{\mathrm{II}} \sin\theta_t^{\mathrm{II}}. \tag{3.42b}$$

From Eqs. (3.41a, 3.41b, 3.41c, 3.41d), the solution for the coefficients is

$$\overline{D}_o = \overline{\overline{C}}^{-1} \overline{D}_i, \tag{3.43}$$

where

$$\overline{D}_o = \begin{pmatrix} D_r^{\mathrm{I}} \\ D_t^{\mathrm{I}} \end{pmatrix}, \tag{3.44a}$$

$$\overline{\overline{C}} = \begin{pmatrix} 1/\varepsilon_1 & -1/\varepsilon_2 \\ \cos\theta_r^{\mathrm{I}}/\sqrt{\varepsilon_1\mu_1} & -\cos\theta_t^{\mathrm{I}}/\sqrt{\varepsilon_2\mu_2} \end{pmatrix}, \tag{3.44b}$$

$$\overline{D}_i = \begin{pmatrix} 1/\varepsilon_1 \\ -\cos\theta_i^{\mathrm{I}}/\sqrt{\varepsilon_1\mu_1} \end{pmatrix} D_i^{\mathrm{I}}, \tag{3.44c}$$

and $D_r^{\mathrm{II}} = D_t^{\mathrm{II}} = 0$. In the conventional material with $\varepsilon > 0$ and $\mu > 0$, only the type-I wave exists for the reflected fields. In the metamaterial with $\varepsilon < 0$ and $\mu < 0$, only the type-I wave exists for the transmitted fields. Besides, since k_2^{I} is positive, negative refraction occurs for the type-I wave, and the refraction angle θ_t^{I} satisfies

$$\sqrt{\varepsilon_1\mu_1} \sin\theta_i^{\mathrm{I}} = \sqrt{\varepsilon_2\mu_2} \sin\theta_t^{\mathrm{I}}, \tag{3.45}$$

as shown in Fig. 3.1. The existence of the solution validates the negative refraction in isotropic metamaterials. It is worth noting that the refraction angle θ_t^{I} is defined to have a positive value, while in Snell's law, θ_t^{I} is defined to be negative and it has a different sign from that of the incident angle, according to Eq. (2.37). So the refractive index of the metamaterial is defined as $n = -\sqrt{\varepsilon_2\mu_2}$; namely, its refractive index is negative.

In applications, people are more interested in the negative refraction that a wave is incident from air. For convenience we use a three-dimensional k space with the axes formed by the components k_x, k_y and k_z to intuitively show that negative refraction can occur at the air–medium interface. We limit our discussion to the incidence of a type-I wave from air to an isotropic medium. The incident plane is the xz plane, and the boundary between the air and the medium is the yz plane. In this geometry the three-dimensional k space reduces to a two-dimensional plane. According to the dispersion relation in Eq. (3.24), the wave vector components satisfy

$$\frac{k_x^2}{\varepsilon_0\mu_0} + \frac{k_z^2}{\varepsilon_0\mu_0} = \omega^2 \tag{3.46a}$$

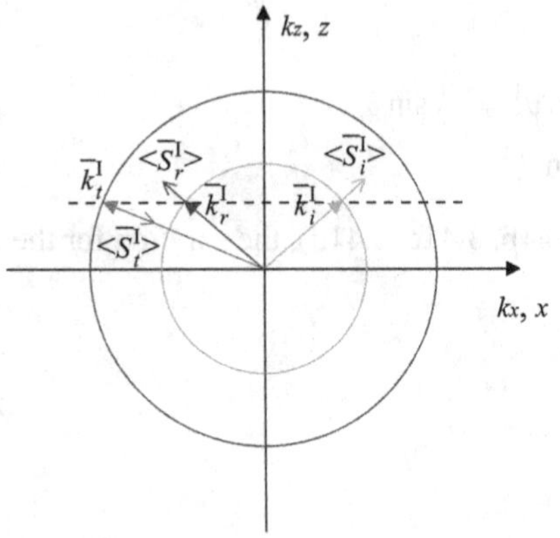

Figure 3.2 k surfaces for the waves in the air and in an isotropic medium, where the uniaxial medium is a medium in case 2 with $\varepsilon < 0$ and $\mu < 0$.

in the air, while in the isotropic medium, according to Eqs. (4.25, 4.27), the wave vector components satisfy

$$\frac{k_x^2}{\varepsilon\mu} + \frac{k_z^2}{\varepsilon\mu} = \omega^2 \tag{3.46b}$$

for both the type-I and type-II waves. We can see that the k surface in the air and in the isotropic media are both circles. In the following we will show one example of negative refraction using the k surface.

The isotropic medium is a medium in case 2 with $\varepsilon < 0$ and $\mu < 0$. In this example, negative refraction is expected to occur for the type-I wave. For the incident field, its wave vector $\vec{k}_i^{\mathrm{I}}$ is in the first quadrant, as shown in Fig. 3.2. According to the continuity condition of the wave vectors in Eq. (3.42a, 3.42b), the wave vector of the reflected field $\vec{k}_r^{\mathrm{I}}$ is in the second quadrant. For the transmitted field, according to Tab. 3.1, the directions of the wave vector and Poynting's vector are $\hat{e}_3$ and $-\hat{e}_3$, respectively. Thus $\vec{k}_t^{\mathrm{I}}$ must be in the second quadrant so that the Poynting's vector is pointing away from the interface, as shown in Fig. 3.2. Negative refraction occurs because of the negative refraction angle of the energy flow.

4 Negative Refraction in Anisotropic Metamaterials

4.1 Introduction

Anisotropic metamaterials have more complicated constitutive parameters, where each of the two constitutive parameter tensors has nine elements, respectively. For simplicity we study the metamaterials where only the diagonal elements in the constitutive parameter tensors are nonzero. Specifically, if two of the three diagonal elements are equal for both the two constitutive parameters, this kind of anisotropic media is a kind of uniaxial media. In this section, we will introduce the basic theoretical concepts and design principles of negative refraction in uniaxial metamaterials as an example.

4.2 Waves in Uniaxial Media

Uniaxial media are a special kind of anisotropic media. The source-free Maxwell equations in uniaxial media in the *xyz* coordinate system are

$$\overline{k} \times \overline{E} = \omega \overline{B}, \tag{4.1a}$$

$$\overline{k} \times \overline{H} = -\omega \overline{D}, \tag{4.1b}$$

$$\overline{k} \cdot \overline{D} = 0, \tag{4.1c}$$

$$\overline{k} \cdot \overline{B} = 0, \tag{4.1d}$$

and the constitutive relations are

$$\overline{D} = \overline{\overline{\varepsilon}} \cdot \overline{E}, \tag{4.2a}$$

$$\overline{B} = \overline{\overline{\mu}} \cdot \overline{H}, \tag{4.2b}$$

in the $\overline{EH}$ representation, where $\overline{\overline{\varepsilon}} = diag\,(\varepsilon, \varepsilon, \varepsilon_z)$ is the permittivity and $\overline{\overline{\mu}} = diag\,(\mu, \mu, \mu_z)$ is the permeability. In the following we will derive the dispersion relations of waves in the uniaxial media in the *kDB* coordinate system.

First, we need to transform the constitutive relations from the $\overline{EH}$ representation in the *xyz* coordinate system to the $\overline{DB}$ representation in the *xyz* coordinate system. According to Eqs. (2.25a, 2.25b, 2.26a, 2.26b, 2.26c, 2.26d, 2.27a, 2.27b, 2.27c, 2.27d), the constitutive relations in the $\overline{DB}$ representation in the *xyz* coordinate system are

$$\overline{E} = \overline{\overline{\kappa}} \cdot \overline{D}, \tag{4.3a}$$

$$\overline{H} = \overline{\overline{\nu}} \cdot \overline{B}, \tag{4.3b}$$

where

$$\overline{\overline{\kappa}} = \overline{\overline{\varepsilon}}^{-1} = diag\,(1/\varepsilon, 1/\varepsilon, 1/\varepsilon_z)\,, \tag{4.4a}$$

$$\overline{\overline{\nu}} = \overline{\overline{\mu}}^{-1} = diag\,(1/\mu, 1/\mu, 1/\mu_z)\,, \tag{4.4b}$$

and

$$\overline{\overline{\varepsilon}} = \overline{\overline{\kappa}}^{-1} = diag\,(1/\kappa, 1/\kappa, 1/\kappa_z)\,, \tag{4.5a}$$

$$\overline{\overline{\mu}} = \overline{\overline{\nu}}^{-1} = diag\,(1/\nu, 1/\nu, 1/\nu_z)\,. \tag{4.5b}$$

Second, we need to transform the Maxwell equations and constitutive relations from the *xyz* coordinate system to the *kDB* coordinate system. From Eqs. (2.32a, 2.32b, 2.32c, 2.32d), the Maxwell equations in the uniaxial media in the *kDB* coordinate system are

$$\overline{k}_k \times \overline{E}_k = \omega \overline{B}_k, \tag{4.6a}$$

$$\overline{k}_k \times \overline{H}_k = -\omega \overline{D}_k, \tag{4.6b}$$

$$\overline{k}_k \cdot \overline{D}_k = 0, \tag{4.6c}$$

$$\overline{k}_k \cdot \overline{B}_k = 0, \tag{4.6d}$$

and the constitutive relations are

$$\overline{E}_k = \overline{\overline{\kappa}}_k \cdot \overline{D}_k, \tag{4.7a}$$

$$\overline{H}_k = \overline{\overline{\nu}}_k \cdot \overline{B}_k, \tag{4.7b}$$

where

$$\overline{\overline{\kappa}}_k = \begin{pmatrix} \kappa & 0 & 0 \\ 0 & \kappa\cos^2\theta + \kappa_z\sin^2\theta & (\kappa - \kappa_z)\sin\theta\cos\theta \\ 0 & (\kappa - \kappa_z)\sin\theta\cos\theta & \kappa\sin^2\theta + \kappa_z\cos^2\theta \end{pmatrix}, \tag{4.8a}$$

$$\overline{\overline{\nu}}_k = \begin{pmatrix} \nu & 0 & 0 \\ 0 & \nu\cos^2\theta + \nu_z\sin^2\theta & (\nu - \nu_z)\sin\theta\cos\theta \\ 0 & (\nu - \nu_z)\sin\theta\cos\theta & \nu\sin^2\theta + \nu_z\cos^2\theta \end{pmatrix}, \tag{4.8b}$$

according to Eqs. (2.28a, 2.28b, 2.29a, 2.29b, 2.29c, 2.29d).

Third, we need to derive the dispersion relations from the coefficient matrix. According to Eqs. (2.35, 2.36a, 2.36b, 2.36c, 2.36d, 4.8a, 4.8b), the determinant of the coefficient matrix is

$$\left| \overline{\overline{\kappa}}'_k - u^2 \overline{\overline{\sigma}}_y \overline{\overline{\nu}}'^{-1}_k \overline{\overline{\sigma}}_y \right| = 0, \tag{4.9}$$

where

$$\overline{\overline{\kappa}}'_k = \begin{pmatrix} \kappa_{11} & 0 \\ 0 & \kappa_{22} \end{pmatrix} = \begin{pmatrix} \kappa & 0 \\ 0 & \kappa\cos^2\theta + \kappa_z\sin^2\theta \end{pmatrix}, \tag{4.10a}$$

$$\overline{\overline{\nu}}'_k = \begin{pmatrix} \nu_{11} & 0 \\ 0 & \nu_{22} \end{pmatrix} = \begin{pmatrix} \nu & 0 \\ 0 & \nu\cos^2\theta + \nu_z\sin^2\theta \end{pmatrix}, \tag{4.10b}$$

$$\overline{\overline{\sigma}}_y = \begin{pmatrix} 0 & -i \\ i & 0 \end{pmatrix}, \tag{4.10c}$$

and $u = \omega/k$. Thus the dispersion relations are

$$u^{\mathrm{I}} = \sqrt{\kappa\left(\nu\cos^2\theta + \nu_z\sin^2\theta\right)}, \tag{4.11a}$$

$$u^{\mathrm{II}} = \sqrt{\left(\kappa \cos^2\theta + \kappa_z \sin^2\theta\right)\nu}. \tag{4.11b}$$

According to Eqs. (2.34a, 2.34b), the two dispersion relations correspond to two types of characteristic waves with $D_1 \neq 0$, $D_2 = 0$ and $D_1 = 0$, $D_2 \neq 0$, respectively. We define the waves with $D_1 \neq 0$ and $D_2 = 0$ as the *type-I waves* and the waves with $D_1 = 0$ and $D_2 \neq 0$ as the *type-II waves*. It is worth noting that in this section we let $\kappa\left(\nu\cos^2\theta + \nu_z\sin^2\theta\right) > 0$ for the type-I waves and $\left(\kappa\cos^2\theta + \kappa_z\sin^2\theta\right)\nu > 0$ for the type-II waves, respectively, in order to study the propagating waves.

Fourth, we study the type-I and type-II waves in the *kDB* coordinate system, respectively. For a type-I wave, the wave vector is

$$\overline{k}_k^{\mathrm{I}} = \hat{e}_3 k^{\mathrm{I}} = \hat{e}_3 \frac{\omega}{\sqrt{\kappa\left(\nu\cos^2\theta + \nu_z\sin^2\theta\right)}}, \tag{4.12}$$

which is always positive. From $D_1 \neq 0$ and $D_2 = 0$, the electric flux density is

$$\overline{D}_k^{\mathrm{I}} = \hat{e}_1 D^{\mathrm{I}}. \tag{4.13a}$$

According to Eq. (2.34a) or Eq. (2.34b), the magnetic flux density is

$$\overline{B}_k^{\mathrm{I}} = \begin{cases} \hat{e}_2\sqrt{\kappa/\bar{\nu}}D^{\mathrm{I}}, & \kappa > 0 \text{ and } \bar{\nu} > 0, \\ -\hat{e}_2\sqrt{\kappa/\bar{\nu}}D^{\mathrm{I}}, & \kappa < 0 \text{ and } \bar{\nu} < 0 \end{cases} \tag{4.13b}$$

where $\bar{\nu} = \nu\cos^2\theta + \nu_z\sin^2\theta$. According to Eq. (4.7a), the electric field is

$$\overline{E}_k^{\mathrm{I}} = \hat{e}_1 \kappa D^{\mathrm{I}}. \tag{4.13c}$$

Then according to Eq. (4.7b), the magnetic field is

$$\overline{H}_k^{\mathrm{I}} = \begin{cases} \hat{e}_2\sqrt{\kappa\bar{\nu}}D^{\mathrm{I}} + \hat{e}_3\left(\nu - \nu_z\right)\sin\theta\cos\theta\sqrt{\kappa/\bar{\nu}}D^{\mathrm{I}}, & \kappa > 0 \text{ and } \bar{\nu} > 0, \\ \hat{e}_2\sqrt{\kappa\bar{\nu}}D^{\mathrm{I}} - \hat{e}_3\left(\nu - \nu_z\right)\sin\theta\cos\theta\sqrt{\kappa/\bar{\nu}}D^{\mathrm{I}}, & \kappa < 0 \text{ and } \bar{\nu} < 0. \end{cases} \tag{4.13d}$$

Similarly, for a type-II wave, the wave vector is

$$\overline{k}_k^{\mathrm{II}} = \hat{e}_3 k^{\mathrm{II}} = \hat{e}_3 \frac{\omega}{\sqrt{\left(\kappa\cos^2\theta + \kappa_z\sin^2\theta\right)\nu}}, \tag{4.14}$$

which is also always positive. The field components are

$$\overline{D}_k^{\mathrm{II}} = \hat{e}_2 D^{\mathrm{II}}, \tag{4.15a}$$

$$\overline{B}_k^{\mathrm{II}} = \begin{cases} -\hat{e}_1\sqrt{\bar{\kappa}/\nu}D^{\mathrm{II}}, & \bar{\kappa} > 0 \text{ and } \nu > 0, \\ \hat{e}_1\sqrt{\bar{\kappa}/\nu}D^{\mathrm{II}}, & \bar{\kappa} < 0 \text{ and } \nu < 0, \end{cases} \tag{4.15b}$$

$$\overline{E}_k^{\mathrm{II}} = \hat{e}_2 \bar{\kappa} D^{\mathrm{II}} + \hat{e}_3 (\kappa - \kappa_z) \sin\theta \cos\theta D^{\mathrm{II}}, \tag{4.15c}$$

$$\overline{H}_k^{\mathrm{II}} = -\hat{e}_1 \sqrt{\bar{\kappa}\nu} D^{\mathrm{II}} \tag{4.15d}$$

where $\bar{\kappa} = \kappa \cos^2\theta + \kappa_z \sin^2\theta$.

The Poynting's vectors in the *kDB* coordinate system for both the type-I and type-II waves can be calculated by

$$\langle \overline{S}_k \rangle = \frac{1}{2} \mathrm{Re} \left(\overline{E}_k \times \overline{H}_k^* \right). \tag{4.16}$$

From Eqs. (4.13c, 4.13d), for a type-I wave, we get

$$\left\langle \overline{S}_k^{\mathrm{I}} \right\rangle = \begin{cases} -\hat{e}_2 \frac{1}{2}\kappa (\nu - \nu_z) \sin\theta \cos\theta \sqrt{\kappa/\bar{\nu}} |D^{\mathrm{I}}|^2 \\ +\hat{e}_3 \frac{1}{2}\kappa \left(\nu \cos^2\theta + \nu_z \sin^2\theta\right) \sqrt{\kappa/\bar{\nu}} |D^{\mathrm{I}}|^2, & \kappa > 0 \text{ and } \bar{\nu} > 0, \\ \hat{e}_2 \frac{1}{2}\kappa (\nu - \nu_z) \sin\theta \cos\theta \sqrt{\kappa/\bar{\nu}} |D^{\mathrm{I}}|^2 \\ -\hat{e}_3 \frac{1}{2}\kappa \left(\nu \cos^2\theta + \nu_z \sin^2\theta\right) \sqrt{\kappa/\bar{\nu}} |D^{\mathrm{I}}|^2, & \kappa < 0 \text{ and } \bar{\nu} < 0. \end{cases} \tag{4.17}$$

Similarly, from Eqs. (4.15c, 4.15d), for a type-II wave, we get

$$\left\langle \overline{S}_k^{\mathrm{II}} \right\rangle = \begin{cases} -\hat{e}_2 \frac{1}{2} (\kappa - \kappa_z) \nu \sin\theta \cos\theta \sqrt{\bar{\kappa}/\nu} |D^{\mathrm{I}}|^2 \\ +\hat{e}_3 \frac{1}{2} \left(\kappa \cos^2\theta + \kappa_z \sin^2\theta\right) \nu \sqrt{\bar{\kappa}/\nu} |D^{\mathrm{I}}|^2, & \bar{\kappa} > 0 \text{ and } \nu > 0, \\ \hat{e}_2 \frac{1}{2} (\kappa - \kappa_z) \nu \sin\theta \cos\theta \sqrt{\bar{\kappa}/\nu} |D^{\mathrm{I}}|^2 \\ -\hat{e}_3 \frac{1}{2} \left(\kappa \cos^2\theta + \kappa_z \sin^2\theta\right) \nu \sqrt{\bar{\kappa}/\nu} |D^{\mathrm{I}}|^2, & \bar{\kappa} < 0 \text{ and } \nu < 0. \end{cases} \tag{4.18}$$

Thus, for the type-I waves, there are two sets of field components, which correspond to $\kappa > 0$, $\nu \cos^2\theta + \nu_z \sin^2\theta > 0$ and $\kappa < 0$, $\nu \cos^2\theta + \nu_z \sin^2\theta < 0$, respectively. For the type-II waves, there are also two sets of field components, which correspond to $\kappa \cos^2\theta + \kappa_z \sin^2\theta > 0$, $\nu > 0$ and $\kappa \cos^2\theta + \kappa_z \sin^2\theta < 0$, $\nu < 0$, respectively. In the following we define the media with $\kappa > 0$, $\nu \cos^2\theta + \nu_z \sin^2\theta > 0$ and $\kappa \cos^2\theta + \kappa_z \sin^2\theta > 0$, $\nu > 0$ as the *conventional materials* and the media with $\kappa < 0$, $\nu \cos^2\theta + \nu_z \sin^2\theta < 0$ and $\kappa \cos^2\theta + \kappa_z \sin^2\theta < 0$, $\nu < 0$ as the *metamaterials*. The type-I and type-II waves constitute a set of eigenstates in uniaxial media, and any type of waves can be expanded using this set of eigenstates.

Finally, we transform the field components from the *kDB* coordinate system back to the *xyz* coordinate system. The coordinate systems are shown in Fig. 2.1. According to Eqs. (2.22a, 2.22b, 2.22c, 2.22d), for the type-I waves, we have

$$\overline{D}^{\mathrm{I}} = \hat{x} \sin\phi D^{\mathrm{I}} - \hat{y} \cos\phi D^{\mathrm{I}}, \tag{4.19a}$$

$$\overline{E}^{\mathrm{I}} = \hat{x} \sin\phi \kappa D^{\mathrm{I}} - \hat{y} \cos\phi \kappa D^{\mathrm{I}}, \tag{4.19b}$$

$$\overline{B}^{\mathrm{I}} = \hat{x}\cos\theta\cos\phi\sqrt{\kappa/\overline{\nu}}D^{\mathrm{I}} + \hat{y}\cos\theta\sin\phi\sqrt{\kappa/\overline{\nu}}D^{\mathrm{I}} - \hat{z}\sin\theta\sqrt{\kappa/\overline{\nu}}D^{\mathrm{I}}, \tag{4.19c}$$

$$\overline{H}^{\mathrm{I}} = \hat{x}\nu\cos\theta\cos\phi\sqrt{\kappa/\overline{\nu}}D^{\mathrm{I}} + \hat{y}\nu\cos\theta\sin\phi\sqrt{\kappa/\overline{\nu}}D^{\mathrm{I}} - \hat{z}\nu_z\sin\theta\sqrt{\kappa/\overline{\nu}}D^{\mathrm{I}}, \tag{4.19d}$$

in the conventional materials, and

$$\overline{D}^{\mathrm{I}} = \hat{x}\sin\phi D^{\mathrm{I}} - \hat{y}\cos\phi D^{\mathrm{I}}, \tag{4.20a}$$

$$\overline{E}^{\mathrm{I}} = \hat{x}\sin\phi\kappa D^{\mathrm{I}} - \hat{y}\cos\phi\kappa D^{\mathrm{I}}, \tag{4.20b}$$

$$\overline{B}^{\mathrm{I}} = -\hat{x}\cos\theta\cos\phi\sqrt{\kappa/\overline{\nu}}D^{\mathrm{I}} - \hat{y}\cos\theta\sin\phi\sqrt{\kappa/\overline{\nu}}D^{\mathrm{I}} + \hat{z}\sin\theta\sqrt{\kappa/\overline{\nu}}D^{\mathrm{I}}, \tag{4.20c}$$

$$\overline{H}^{\mathrm{I}} = -\hat{x}\nu\cos\theta\cos\phi\sqrt{\kappa/\overline{\nu}}D^{\mathrm{I}} - \hat{y}\nu\cos\theta\sin\phi\sqrt{\kappa/\overline{\nu}}D^{\mathrm{I}} + \hat{z}\nu_z\sin\theta\sqrt{\kappa/\overline{\nu}}D^{\mathrm{I}}, \tag{4.20d}$$

in the metamaterials, respectively. Similarly, for the type-II waves, we have

$$\overline{D}_k^{\mathrm{II}} = \hat{x}\cos\theta\cos\phi D^{\mathrm{II}} + \hat{y}\cos\theta\sin\phi D^{\mathrm{II}} - \hat{z}\sin\theta D^{\mathrm{II}}, \tag{4.21a}$$

$$\overline{E}_k^{\mathrm{II}} = \hat{x}\cos\theta\cos\phi\kappa D^{\mathrm{II}} + \hat{y}\cos\theta\sin\phi\kappa D^{\mathrm{II}} - \hat{z}\sin\theta\kappa_z D^{\mathrm{II}}, \tag{4.21b}$$

$$\overline{B}_k^{\mathrm{II}} = -\hat{x}\sin\phi\sqrt{\overline{\kappa}/\nu}D^{\mathrm{II}} + \hat{y}\cos\phi\sqrt{\overline{\kappa}/\nu}D^{\mathrm{II}}, \tag{4.21c}$$

$$\overline{H}_k^{\mathrm{II}} = -\hat{x}\sin\phi\sqrt{\overline{\kappa}\nu}D^{\mathrm{II}} + \hat{y}\cos\phi\sqrt{\overline{\kappa}\nu}D^{\mathrm{II}}, \tag{4.21d}$$

in the conventional materials, and

$$\overline{D}_k^{\mathrm{II}} = \hat{x}\cos\theta\cos\phi D^{\mathrm{II}} + \hat{y}\cos\theta\sin\phi D^{\mathrm{II}} - \hat{z}\sin\theta D^{\mathrm{II}}, \tag{4.22a}$$

$$\overline{E}_k^{\mathrm{II}} = \hat{x}\cos\theta\cos\phi\kappa D^{\mathrm{II}} + \hat{y}\cos\theta\sin\phi\kappa D^{\mathrm{II}} - \hat{z}\sin\theta\kappa_z D^{\mathrm{II}}, \tag{4.22b}$$

$$\overline{B}_k^{\mathrm{II}} = \hat{x}\sin\phi\sqrt{\overline{\kappa}/\nu}D^{\mathrm{II}} - \hat{y}\cos\phi\sqrt{\overline{\kappa}/\nu}D^{\mathrm{II}}, \tag{4.22c}$$

$$\overline{H}_k^{\mathrm{II}} = -\hat{x}\sin\phi\sqrt{\overline{\kappa}\nu}D^{\mathrm{II}} + \hat{y}\cos\phi\sqrt{\overline{\kappa}\nu}D^{\mathrm{II}}, \tag{4.22d}$$

in the metamaterials, respectively. Note that the above field components are expressed in the $\overline{DB}$ representation.

Transforming from the $\overline{DB}$ representation to the $\overline{EH}$ representation, the dispersion relations are

$$k^{\mathrm{I}} = \omega\sqrt{\frac{\varepsilon\mu\mu_z}{\overline{\mu}}}, \tag{4.23a}$$

$$k^{\mathrm{II}} = \omega\sqrt{\frac{\varepsilon\varepsilon_z\mu}{\overline{\varepsilon}}}, \tag{4.23b}$$

according to Eqs. (4.4a, 4.4b, 4.11a, 4.11b), where

$$\overline{\mu} = \mu_z\cos^2\theta + \mu\sin^2\theta, \tag{4.24a}$$

$$\overline{\varepsilon} = \varepsilon_z\cos^2\theta + \varepsilon\sin^2\theta. \tag{4.24b}$$

We get four cases in total. For case 1 the conventional materials with $\kappa > 0$ and $\nu \cos^2\theta + \nu_z \sin^2\theta > 0$ require that $\varepsilon > 0$ and $(1/\mu)\cos^2\theta + (1/\mu_z)\sin^2\theta > 0$. For case 2 the metamaterials with $\kappa < 0$ and $\nu \cos^2\theta + \nu_z \sin^2\theta < 0$ require that $\varepsilon < 0$ and $(1/\mu)\cos^2\theta + (1/\mu_z)\sin^2\theta < 0$. For case 3 the conventional materials with $\kappa \cos^2\theta + \kappa_z \sin^2\theta > 0$ and $\nu > 0$ require that $(1/\varepsilon)\cos^2\theta + (1/\varepsilon_z)\sin^2\theta > 0$ and $\mu > 0$. For case 4 the metamaterials with $\kappa \cos^2\theta + \kappa_z \sin^2\theta < 0$ and $\nu < 0$ require that $(1/\varepsilon)\cos^2\theta + (1/\varepsilon_z)\sin^2\theta < 0$ and $\mu < 0$. Case 1 and case 2 correspond to the type-I waves, and case 3 and case 4 correspond to the type-II waves. For a type-I wave, the wave number is

$$k^{\mathrm{I}} = \omega\sqrt{\frac{\varepsilon\mu\mu_z}{\mu_z\cos^2\theta + \mu\sin^2\theta}}, \tag{4.25}$$

which is always positive. For brevity the field components are omitted here. Readers can easily derive them by themselves. Since $\bar{k} = \hat{e}_3 k = \hat{x}k\sin\theta\cos\phi + \hat{y}k\sin\theta\sin\phi + \hat{z}k\cos\theta$ and $\bar{r} = \hat{x}x + \hat{y}y + \hat{z}z$, the field components contain a phase factor, where

$$D^{\mathrm{I}} = D_0^{\mathrm{I}} e^{i\bar{k}^{\mathrm{I}}\cdot\bar{r}} = D_0^{\mathrm{I}} e^{ik^{\mathrm{I}}(x\sin\theta\cos\phi + y\sin\theta\sin\phi + z\cos\theta)}. \tag{4.26}$$

Similarly, for the type II wave, the wave number is

$$k^{\mathrm{II}} = \omega\sqrt{\frac{\varepsilon\varepsilon_z\mu}{\varepsilon_z\cos^2\theta + \varepsilon\sin^2\theta}}, \tag{4.27}$$

which is also always positive. The field components are also omitted here, and they contain a phase factor, where

$$D^{\mathrm{II}} = D_0^{\mathrm{II}} e^{i\bar{k}^{\mathrm{II}}\cdot\bar{r}} = D_0^{\mathrm{II}} e^{ik^{\mathrm{II}}(x\sin\theta\cos\phi + y\sin\theta\sin\phi + z\cos\theta)}. \tag{4.28}$$

The Poynting's vectors can be calculated by

$$\langle\bar{S}\rangle = \frac{1}{2}\mathrm{Re}\left(\bar{E}\times\bar{H}^*\right). \tag{4.29}$$

For the type-I waves, we get

$$\left\langle\bar{S}^{\mathrm{I}}\right\rangle = \begin{cases} \hat{e}_3'\frac{1}{2\varepsilon\mu_z}\sqrt{\frac{\mu\mu_z}{\varepsilon\bar{\mu}}}D_0^{\mathrm{I}2}, & \varepsilon > 0 \text{ and } \cos^2\theta/\mu + \sin^2\theta/\mu_z > 0, \\ -\hat{e}_3'\frac{1}{2\varepsilon\mu_z}\sqrt{\frac{\mu\mu_z}{\varepsilon\bar{\mu}}}D_0^{\mathrm{I}2}, & \varepsilon < 0 \text{ and } \cos^2\theta/\mu + \sin^2\theta/\mu_z < 0, \end{cases} \tag{4.30}$$

where $\hat{e}_3' = \hat{x}\sin\theta\cos\phi + \hat{y}\sin\theta\sin\phi + \hat{z}\cos\theta\mu_z/\mu$ and $\bar{\mu} = \mu_z\cos^2\theta + \mu\sin^2\theta$. Note that $\hat{e}_3'$ is different from $\hat{e}_3$. Similarly, for the type-II waves, we get

Table 4.1 The main properties of type-I waves in uniaxial media.

type-I wave	**direction of $\overline{\mathbf{k}}^{\mathrm{I}}$**	**direction of $\left\langle \overline{\mathbf{S}}^{\mathrm{I}} \right\rangle$**
case 1: $\varepsilon > 0$ and $\cos^2\theta/\mu + \sin^2\theta/\mu_z > 0$	$\hat{e}_3$	$\operatorname{sgn}(\varepsilon\mu_z)\,\hat{e}'_3$
case 2: $\varepsilon < 0$ and $\cos^2\theta/\mu + \sin^2\theta/\mu_z < 0$	$\hat{e}_3$	$-\operatorname{sgn}(\varepsilon\mu_z)\,\hat{e}'_3$

Table 4.2 The main properties of type-II waves in uniaxial media.

type-II wave	**direction of $\overline{\mathbf{k}}^{\mathrm{II}}$**	**direction of $\left\langle \overline{\mathbf{S}}^{\mathrm{II}} \right\rangle$**
case 3: $\cos^2\theta/\varepsilon + \sin^2\theta/\varepsilon_z > 0$ and $\mu > 0$	$\hat{e}_3$	$\operatorname{sgn}(\varepsilon_z\mu)\,\hat{e}'_3$
case 4: $\cos^2\theta/\varepsilon + \sin^2\theta/\varepsilon_z < 0$ and $\mu < 0$	$\hat{e}_3$	$-\operatorname{sgn}(\varepsilon_z\mu)\,\hat{e}'_3$

$$\left\langle \overline{S}^{\mathrm{II}} \right\rangle = \begin{cases} \hat{e}'_3 \frac{1}{2\varepsilon_z\mu}\sqrt{\frac{\mu\bar{\varepsilon}}{\varepsilon\varepsilon_z}}D_0^{\mathrm{II}2}, & \cos^2\theta/\varepsilon + \sin^2\theta/\varepsilon_z > 0 \text{ and } \mu > 0, \\ -\hat{e}'_3 \frac{1}{2\varepsilon_z\mu}\sqrt{\frac{\mu\bar{\varepsilon}}{\varepsilon\varepsilon_z}}D_0^{\mathrm{II}2}, & \cos^2\theta/\varepsilon + \sin^2\theta/\varepsilon_z < 0 \text{ and } \mu < 0, \end{cases} \tag{4.31}$$

where $\hat{e}'_3 = \hat{x}\sin\theta\cos\phi + \hat{y}\sin\theta\sin\phi + \hat{z}\cos\theta\varepsilon_z/\varepsilon$ and $\bar{\varepsilon} = \varepsilon_z\cos^2\theta + \varepsilon\sin^2\theta$. In light of the directions of the wave vectors and Poynting's vectors, the main properties of type-I and type-II waves are summarized in Tabs. 4.1 and 4.2.

At the end of this section we would like to compare the waves in isotropic and uniaxial metamaterials. In either of the two kinds of media, there are two types of waves: type-I and type-II waves. For any type of waves, it corresponds to two types of media: conventional materials and metamaterials. Besides these similarities, there is one important difference between the two kinds of metamaterials. The directions of the wave vector and Poynting's vector are always the same in isotropic metamaterials, while in uniaxial metamaterials generally they are not in the same direction.

4.3 Negative Refraction in Uniaxial Metamaterials

In this section we show that negative refraction can occur at an interface between two uniaxial media. We let the incident plane be the *xz* plane and the boundary between the two media be the *yz* plane, as shown in Fig. 4.1. All the angles are assumed to have positive values.

As an example, we only consider the incidence of a type-I wave and its refraction from a medium in case 1 to a medium in case 2. Specifically, we let $\varepsilon_z > 0$, $\mu > 0$ and $\mu_z > 0$ in case 1, and $\varepsilon_z < 0$, $\mu < 0$, $\mu_z < 0$ in case 2, so

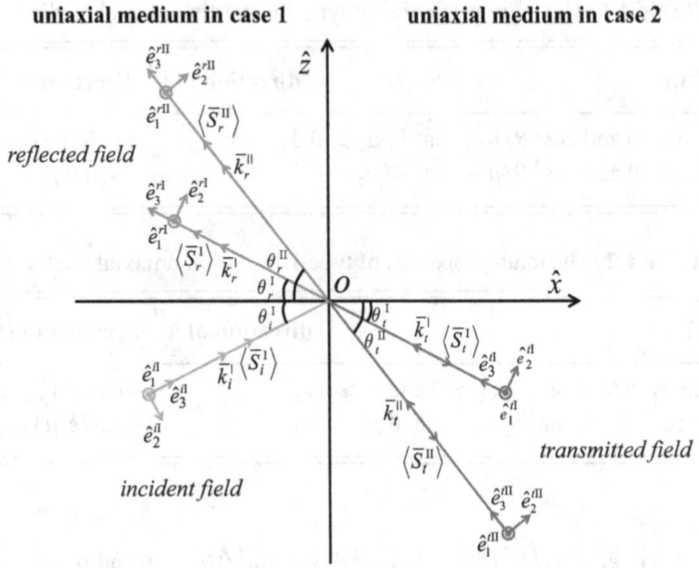

Figure 4.1 A schematic of the negative refraction at an interface between a uniaxial medium in case 1 and a uniaxial medium in case 2.

that the medium in case 1 is a medium in case 3 and the medium in case 2 is a medium in case 4 for the type-II waves. The field components and Poynting's vectors for both the type-I and type-II waves in the two media were derived at the end of the last section. For the incident field, $\theta = \pi/2 - \theta_i^{\mathrm{I}}$ and $\phi = 0$. Then the incident field components are

$$\overline{D}_i^{\mathrm{I}} = -\hat{y} D_i^{\mathrm{I}} e^{ik_i^{\mathrm{I}}\left(x\cos\theta_i^{\mathrm{I}} + z\sin\theta_i^{\mathrm{I}}\right)}, \tag{4.32a}$$

$$\overline{E}_i^{\mathrm{I}} = -\hat{y}\frac{1}{\varepsilon_1} D_i^{\mathrm{I}} e^{ik_i^{\mathrm{I}}\left(x\cos\theta_i^{\mathrm{I}} + z\sin\theta_i^{\mathrm{I}}\right)}, \tag{4.32b}$$

$$\overline{B}_i^{\mathrm{I}} = \left(\hat{x}\sin\theta_i^{\mathrm{I}} - \hat{z}\cos\theta_i^{\mathrm{I}}\right)\eta_i^{\mathrm{I}} D_i^{\mathrm{I}} e^{ik_i^{\mathrm{I}}\left(x\cos\theta_i^{\mathrm{I}} + z\sin\theta_i^{\mathrm{I}}\right)}, \tag{4.32c}$$

$$\overline{H}_i^{\mathrm{I}} = \left(\hat{x}\frac{\sin\theta_i^{\mathrm{I}}}{\mu_1} - \hat{z}\frac{\cos\theta_i^{\mathrm{I}}}{\mu_{z_1}}\right)\eta_i^{\mathrm{I}} D_i^{\mathrm{I}} e^{ik_i^{\mathrm{I}}\left(x\cos\theta_i^{\mathrm{I}} + z\sin\theta_i^{\mathrm{I}}\right)}, \tag{4.32d}$$

and the Poynting's vector is

$$\left\langle \overline{S}_i^{\mathrm{I}} \right\rangle = \left(\hat{x}\cos\theta_i^{\mathrm{I}} + \hat{z}\sin\theta_i^{\mathrm{I}}\frac{\mu_{z_1}}{\mu_1}\right)\frac{\eta_i^{\mathrm{I}}}{2\varepsilon_1\mu_{z_1}} D_i^{\mathrm{I}2}, \tag{4.33}$$

where $\eta_i^{\mathrm{I}} = \sqrt{\mu_1\mu_{z_1}/\left[\varepsilon_1\left(\mu_{z_1}\sin^2\theta_i^{\mathrm{I}} + \mu_1\cos^2\theta_i^{\mathrm{I}}\right)\right]}$ is the impedance and $k_i^{\mathrm{I}} = \omega\sqrt{\varepsilon_1\mu_1\mu_{z_1}/\left(\mu_{z_1}\sin^2\theta_i^{\mathrm{I}} + \mu_1\cos^2\theta_i^{\mathrm{I}}\right)}$ is positive. Considering the direction of the Poynting's vector, we have $\mu_1 > 0$ and $\mu_{z_1} > 0$. For the reflected fields, $\theta = \pi/2 - \theta_r^{\mathrm{I}}$ and $\phi = \pi$ for the type-I wave; and $\theta = \pi/2 - \theta_r^{\mathrm{II}}$

and $\phi = \pi$ for the type-II wave. Then the reflected field components for the type-I wave are

$$\overline{D}_r^{\mathrm{I}} = \hat{y} D_r^{\mathrm{I}} e^{ik_r^{\mathrm{I}}\left(-x\cos\theta_r^{\mathrm{I}} + z\sin\theta_r^{\mathrm{I}}\right)}, \tag{4.34a}$$

$$\overline{E}_r^{\mathrm{I}} = \hat{y}\frac{1}{\varepsilon_1} D_r^{\mathrm{I}} e^{ik_r^{\mathrm{I}}\left(-x\cos\theta_r^{\mathrm{I}} + z\sin\theta_r^{\mathrm{I}}\right)}, \tag{4.34b}$$

$$\overline{B}_r^{\mathrm{I}} = \left(-\hat{x}\sin\theta_r^{\mathrm{I}} - \hat{z}\cos\theta_r^{\mathrm{I}}\right) \eta_r^{\mathrm{I}} D_r^{\mathrm{I}} e^{ik_r^{\mathrm{I}}\left(-x\cos\theta_r^{\mathrm{I}} + z\sin\theta_r^{\mathrm{I}}\right)}, \tag{4.34c}$$

$$\overline{H}_r^{\mathrm{I}} = \left(-\hat{x}\frac{\sin\theta_r^{\mathrm{I}}}{\mu_1} - \hat{z}\frac{\cos\theta_r^{\mathrm{I}}}{\mu_{z_1}}\right) \eta_r^{\mathrm{I}} D_r^{\mathrm{I}} e^{ik_r^{\mathrm{I}}\left(-x\cos\theta_r^{\mathrm{I}} + z\sin\theta_r^{\mathrm{I}}\right)}, \tag{4.34d}$$

and the Poynting's vector is

$$\left\langle \overline{S}_r^{\mathrm{I}} \right\rangle = \left(-\hat{x}\cos\theta_r^{\mathrm{I}} + \hat{z}\sin\theta_r^{\mathrm{I}}\frac{\mu_{z_1}}{\mu_1}\right) \frac{\eta_r^{\mathrm{I}}}{2\varepsilon_1\mu_{z_1}} D_r^{\mathrm{I}2}, \tag{4.35}$$

where $\eta_r^{\mathrm{I}} = \sqrt{\mu_1\mu_{z_1} / \left[\varepsilon_1\left(\mu_{z_1}\sin^2\theta_r^{\mathrm{I}} + \mu_1\cos^2\theta_r^{\mathrm{I}}\right)\right]}$ is the impedance and $k_r^{\mathrm{I}} = \omega\sqrt{\varepsilon_1\mu_1\mu_{z_1} / \left(\mu_{z_1}\sin^2\theta_r^{\mathrm{I}} + \mu_1\cos^2\theta_r^{\mathrm{I}}\right)}$ is positive. Similarly, the reflected field components for the type-II wave are

$$\overline{D}_r^{\mathrm{II}} = \left(-\hat{x}\sin\theta_r^{\mathrm{II}} - \hat{z}\cos\theta_r^{\mathrm{II}}\right) D_r^{\mathrm{II}} e^{ik_r^{\mathrm{II}}\left(-x\cos\theta_r^{\mathrm{II}} + z\sin\theta_r^{\mathrm{II}}\right)}, \tag{4.36a}$$

$$\overline{E}_r^{\mathrm{II}} = \left(-\hat{x}\frac{\sin\theta_r^{\mathrm{II}}}{\varepsilon_1} - \hat{z}\frac{\cos\theta_r^{\mathrm{II}}}{\varepsilon_{z_1}}\right) D_r^{\mathrm{II}} e^{ik_r^{\mathrm{II}}\left(-x\cos\theta_r^{\mathrm{II}} + z\sin\theta_r^{\mathrm{II}}\right)}, \tag{4.36b}$$

$$\overline{B}_r^{\mathrm{II}} = -\hat{y}\eta_r^{\mathrm{II}} D_r^{\mathrm{II}} e^{ik_r^{\mathrm{II}}\left(-x\cos\theta_r^{\mathrm{II}} + z\sin\theta_r^{\mathrm{II}}\right)}, \tag{4.36c}$$

$$\overline{H}_r^{\mathrm{II}} = -\hat{y}\frac{1}{\mu_1}\eta_r^{\mathrm{II}} D_r^{\mathrm{II}} e^{ik_r^{\mathrm{II}}\left(-x\cos\theta_r^{\mathrm{II}} + z\sin\theta_r^{\mathrm{II}}\right)}, \tag{4.36d}$$

and the Poynting's vector is

$$\left\langle \overline{S}_r^{\mathrm{II}} \right\rangle = \left(-\hat{x}\cos\theta_r^{\mathrm{II}} + \hat{z}\cos\theta_r^{\mathrm{II}}\frac{\varepsilon_{z_1}}{\varepsilon_1}\right) \frac{\eta_r^{\mathrm{II}}}{2\varepsilon_{z_1}\mu_1} D_r^{\mathrm{II}2}, \tag{4.37}$$

where $\eta_r^{\mathrm{II}} = \sqrt{\mu_1\left(\varepsilon_{z_1}\sin^2\theta_r^{\mathrm{II}} + \varepsilon_1\cos^2\theta_r^{\mathrm{II}}\right) / \left(\varepsilon_1\varepsilon_{z_1}\right)}$ is the impedance and $k_r^{\mathrm{II}} = \omega\sqrt{\varepsilon_1\varepsilon_{z_1}\mu_1 / \left(\varepsilon_{z_1}\sin^2\theta_r^{\mathrm{II}} + \varepsilon_1\cos^2\theta_r^{\mathrm{II}}\right)}$ is positive. Considering the direction of the Poynting's vector, we have $\varepsilon_1 > 0$ and $\varepsilon_{z_1} > 0$. For the transmitted fields, $\theta = \pi/2 - \theta_t^{\mathrm{I}}$ and $\phi = \pi$ for the type-I wave; and $\theta = \pi/2 - \theta_t^{\mathrm{II}}$ and $\phi = \pi$ for the type-II wave by considering the directions of the energy flows. Then the transmitted field components for the type-I wave are

$$\overline{D}_t^{\mathrm{I}} = \hat{y} D_t^{\mathrm{I}} e^{ik_t^{\mathrm{I}}\left(-x\cos\theta_t^{\mathrm{I}} + z\sin\theta_t^{\mathrm{I}}\right)}, \tag{4.38a}$$

$$\overline{E}_t^{\mathrm{I}} = \hat{y}\frac{1}{\varepsilon_2} D_t^{\mathrm{I}} e^{ik_t^{\mathrm{I}}\left(-x\cos\theta_t^{\mathrm{I}} + z\sin\theta_t^{\mathrm{I}}\right)}, \tag{4.38b}$$

$$\overline{B}_t^{\mathrm{I}} = \left(\hat{x}\sin\theta_t^{\mathrm{I}} + \hat{z}\cos\theta_t^{\mathrm{I}}\right)\eta_t^{\mathrm{I}} D_t^{\mathrm{I}} e^{ik_t^{\mathrm{I}}\left(-x\cos\theta_t^{\mathrm{I}} + z\sin\theta_t^{\mathrm{I}}\right)}, \tag{4.38c}$$

$$\overline{H}_t^{\mathrm{I}} = \left(\hat{x}\frac{\sin\theta_t^{\mathrm{I}}}{\mu_2} + \hat{z}\frac{\cos\theta_t^{\mathrm{I}}}{\mu_{z2}}\right)\eta_t^{\mathrm{I}} D_t^{\mathrm{I}} e^{ik_t^{\mathrm{I}}\left(-x\cos\theta_t^{\mathrm{I}} + z\sin\theta_t^{\mathrm{I}}\right)}, \tag{4.38d}$$

and the Poynting's vector is

$$\left\langle \overline{S}_t^{\mathrm{I}} \right\rangle = -\left(-\hat{x}\cos\theta_t^{\mathrm{I}} + \hat{z}\sin\theta_t^{\mathrm{I}}\frac{\mu_{z2}}{\mu_2}\right)\frac{\eta_t^{\mathrm{I}}}{2\varepsilon_2\mu_{z_2}} D_t^{\mathrm{I}2}, \tag{4.39}$$

where $\eta_t^{\mathrm{I}} = \sqrt{\mu_2\mu_{z_2}/\left[\varepsilon_2\left(\mu_{z_2}\sin^2\theta_t^{\mathrm{I}} + \mu_2\cos^2\theta_t^{\mathrm{I}}\right)\right]}$ is the impedance and $k_t^{\mathrm{I}} = \omega\sqrt{\varepsilon_2\mu_2\mu_{z_2}/\left(\mu_{z_2}\sin^2\theta_t^{\mathrm{I}} + \mu_2\cos^2\theta_t^{\mathrm{I}}\right)}$ is positive. Considering the direction of the Poynting's vector, we have $\mu_2 < 0$ and $\mu_{z_2} < 0$. Similarly, the transmitted field components for the type-II wave are

$$\overline{D}_t^{\mathrm{II}} = \left(-\hat{x}\sin\theta_t^{\mathrm{II}} - \hat{z}\cos\theta_t^{\mathrm{II}}\right) D_t^{\mathrm{II}} e^{ik_2^{\mathrm{II}}\left(-x\cos\theta_t^{\mathrm{II}} + z\sin\theta_t^{\mathrm{II}}\right)}, \tag{4.40a}$$

$$\overline{E}_t^{\mathrm{II}} = \left(-\hat{x}\sin\theta_t^{\mathrm{II}}\frac{1}{\varepsilon_2} - \hat{z}\cos\theta_t^{\mathrm{II}}\frac{1}{\varepsilon_{z2}}\right) D_t^{\mathrm{II}} e^{ik_2^{\mathrm{II}}\left(-x\cos\theta_t^{\mathrm{II}} + z\sin\theta_t^{\mathrm{II}}\right)}, \tag{4.40b}$$

$$\overline{B}_t^{\mathrm{II}} = \hat{y}\eta_t^{\mathrm{II}} D_t^{\mathrm{II}} e^{ik_2^{\mathrm{II}}\left(-x\cos\theta_t^{\mathrm{II}} + z\sin\theta_t^{\mathrm{II}}\right)}, \tag{4.40c}$$

$$\overline{H}_t^{\mathrm{II}} = \hat{y}\frac{1}{\mu_2}\eta_t^{\mathrm{II}} D_t^{\mathrm{II}} e^{ik_2^{\mathrm{II}}\left(-x\cos\theta_t^{\mathrm{II}} + z\sin\theta_t^{\mathrm{II}}\right)}, \tag{4.40d}$$

and the Poynting's vector is

$$\left\langle \overline{S}_t^{\mathrm{II}} \right\rangle = -\left(-\hat{x}\cos\theta_t^{\mathrm{II}} + \hat{z}\sin\theta_t^{\mathrm{II}}\frac{\varepsilon_{z2}}{\varepsilon_2}\right)\frac{\eta_t^{\mathrm{II}}}{2\varepsilon_{z_2}\mu_2} D_t^{\mathrm{II}2}, \tag{4.41}$$

where $\eta_t^{\mathrm{II}} = \sqrt{\mu_2\left(\varepsilon_{z_2}\sin^2\theta_t^{\mathrm{II}} + \varepsilon_2\cos^2\theta_t^{\mathrm{II}}\right)/\left(\varepsilon_2\varepsilon_{z_2}\right)}$ is the impedance and $k_t^{\mathrm{II}} = \omega\sqrt{\varepsilon_2\varepsilon_{z_2}\mu_2/\left(\varepsilon_{z_2}\sin^2\theta_t^{\mathrm{II}} + \varepsilon_2\cos^2\theta_t^{\mathrm{II}}\right)}$ is positive. Considering the direction of the Poynting's vector, we have $\varepsilon_2 < 0$ and $\varepsilon_{z_2} < 0$. Note that the direction of the energy flow is always from the interface to infinity.

Considering the continuity conditions for the electric fields and magnetic fields, we get

$$-\frac{1}{\varepsilon_1}D_i^{\mathrm{I}} + \frac{1}{\varepsilon_1}D_r^{\mathrm{I}} = \frac{1}{\varepsilon_2}D_t^{\mathrm{I}}, \tag{4.42a}$$

$$\frac{\cos\theta_r^{\mathrm{II}}}{\varepsilon_{z1}}D_r^{\mathrm{II}} = \frac{\cos\theta_t^{\mathrm{II}}}{\varepsilon_{z2}}D_t^{\mathrm{II}}, \tag{4.42b}$$

$$\frac{1}{\mu_1}\eta_r^{\mathrm{II}}D_r^{\mathrm{II}} = -\frac{1}{\mu_2}\eta_t^{\mathrm{II}}D_t^{\mathrm{II}}, \tag{4.42c}$$

$$\frac{\cos\theta_i^{\mathrm{I}}}{\mu_{z_1}}\eta_i^{\mathrm{I}}D_i^{\mathrm{I}} + \frac{\cos\theta_r^{\mathrm{I}}}{\mu_{z_1}}\eta_r^{\mathrm{I}}D_r^{\mathrm{I}} = -\frac{\cos\theta_t^{\mathrm{I}}}{\mu_{z_2}}\eta_t^{\mathrm{I}}D_t^{\mathrm{I}}, \tag{4.42d}$$

and

$$k_i^{\mathrm{I}} \sin\theta_i^{\mathrm{I}} = k_r^{\mathrm{I}} \sin\theta_r^{\mathrm{I}} = k_t^{\mathrm{I}} \sin\theta_t^{\mathrm{I}}, \tag{4.43a}$$

$$k_r^{\mathrm{II}} \sin\theta_r^{\mathrm{II}} = k_t^{\mathrm{II}} \sin\theta_t^{\mathrm{II}}. \tag{4.43b}$$

The latter two equations are the continuity conditions of the wave vectors for the type-I and type-II waves, respectively. From Eqs. (4.42a, 4.42b, 4.42c, 4.42d), the solution for the coefficients is

$$\overline{D}_o = \overline{\overline{C}}^{-1} \overline{D}_i, \tag{4.44}$$

where

$$\overline{D}_o = \begin{pmatrix} D_r^{\mathrm{I}} \\ D_t^{\mathrm{I}} \end{pmatrix}, \tag{4.45a}$$

$$\overline{\overline{C}} = \begin{pmatrix} 1/\varepsilon_1 & -1/\varepsilon_2 \\ \cos\theta_r^{\mathrm{I}}\eta_r^{\mathrm{I}}/\mu_{z1} & -\cos\theta_t^{\mathrm{I}}\eta_t^{\mathrm{I}}/\mu_{z2} \end{pmatrix}, \tag{4.45b}$$

$$\overline{D}_i = \begin{pmatrix} 1/\varepsilon_1 \\ -\cos\theta_i^{\mathrm{I}}\eta_i^{\mathrm{I}}/\mu_{z1} \end{pmatrix} D_i^{\mathrm{I}}, \tag{4.45c}$$

and $D_r^{\mathrm{II}} = D_t^{\mathrm{II}} = 0$. In the medium in case 1 with $\varepsilon > 0$ and $(1/\mu)\cos^2\theta + (1/\mu_z)\sin^2\theta > 0$, only the type-I wave exists for the reflected fields. In the medium in case 2 with $\varepsilon > 0$ and $(1/\mu)\cos^2\theta + (1/\mu_z)\sin^2\theta < 0$, only the type-I wave exists for the transmitted fields. Besides, since k_t^{I} is positive, negative refraction occurs for the type-I wave and the refraction angle θ_t^{I} satisfies

$$\sqrt{\frac{\varepsilon_1\mu_1\mu_{z1}}{\mu_{z1}\sin^2\theta_i^{\mathrm{I}} + \mu_1\cos^2\theta_i^{\mathrm{I}}}}\sin\theta_i^{\mathrm{I}} = \sqrt{\frac{\varepsilon_2\mu_2\mu_{z2}}{\mu_{z2}\sin^2\theta_t^{\mathrm{I}} + \mu_2\cos^2\theta_t^{\mathrm{I}}}}\sin\theta_t^{\mathrm{I}}, \tag{4.46}$$

as shown in Fig. 4.1. The existence of the solution validates the negative refraction in uniaxial metamaterials.

In applications people are more interested in the negative refraction that a wave is incident from air. For convenience we use the *k* surface to intuitively show that negative refraction can occur at the air–medium interface. We limit our discussion to the incidence of a type-I wave from air to a uniaxial medium. The incident plane is the *xz* plane, and the boundary between the air and the medium is the *yz* plane. In this geometry the three-dimensional *k* space reduces to a two-dimensional plane. According to the dispersion relation in Eq. (3.24), the wave vector components satisfy

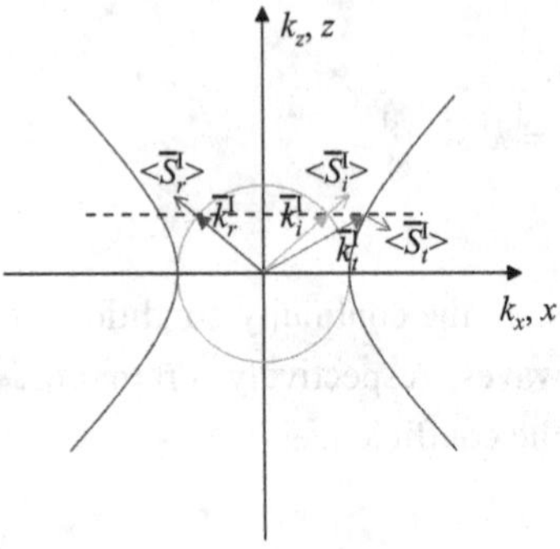

Figure 4.2 k surfaces for the waves in the air and in a uniaxial medium, where the uniaxial medium is a medium in case 1 with $\varepsilon > 0$, $\mu < 0$ and $\mu_z > 0$.

$$\frac{k_x^2}{\varepsilon_0\mu_0} + \frac{k_z^2}{\varepsilon_0\mu_0} = \omega^2 \tag{4.47a}$$

in the air, while in the uniaxial medium, according to Eqs. (4.25, 4.27), the wave vector components satisfy

$$\frac{k_x^2}{\varepsilon\mu_z} + \frac{k_z^2}{\varepsilon\mu} = \omega^2 \tag{4.47b}$$

for a type-I wave, and

$$\frac{k_x^2}{\varepsilon\mu_z} + \frac{k_z^2}{\varepsilon\mu} = \omega^2 \tag{4.47c}$$

for a type-II wave. We can see that the k surface in the air is always a circle, while the k surface in the uniaxial medium is an ellipse or hyperbola. In the following we will show some examples of negative refraction using the k surface.

1. The uniaxial medium is a medium in case 1 with $\varepsilon > 0$, $\mu < 0$ and $\mu_z > 0$. In this example, negative refraction is expected to occur for the type-I wave. For the incident field, its wave vector k_i^{I} is in the first quadrant, as shown in Fig. 4.2. According to the continuity condition of the wave vectors in Eq. (4.43a, 4.43b), the wave vector of the reflected field k_r^{I} is in the second quadrant. For the transmitted field, according to Tab. 4.1, the directions of the wave vector and Poynting's vector are $\hat{e}_3$ and $\hat{e}_3'$, respectively. Thus k_t^{I} must be in the first quadrant so that the Poynting's vector is pointing away from the interface, as shown in Fig. 4.2. Negative refraction occurs because of the negative refraction angle of the energy flow.

2. The uniaxial medium is a medium in case 2 with $\varepsilon < 0$, $\mu < 0$ and $\mu_z < 0$. In this example, negative refraction is also expected to occur for the type-I

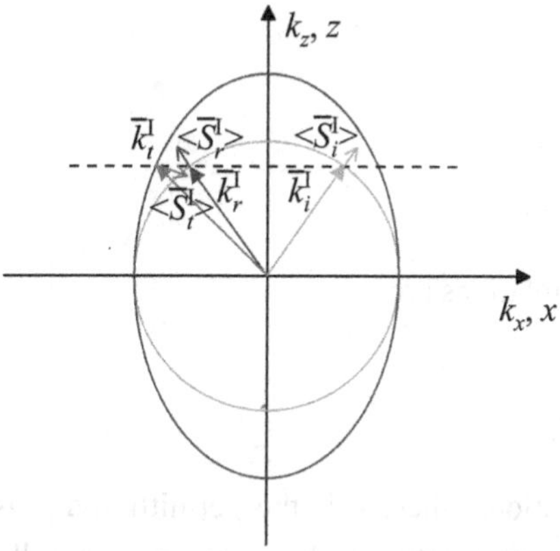

Figure 4.3 k surfaces for the waves in the air and in a uniaxial medium, where the uniaxial medium is a medium in case 2 with $\varepsilon < 0$, $\mu < 0$ and $\mu_z < 0$.

wave. For the incident field, its wave vector k_i^{I} is in the first quadrant, as shown in Fig. 4.3. According to the continuity condition of the wave vectors, the wave vector of the reflected field k_r^{I} is in the second quadrant. For the transmitted field, according to Tab. 4.1, the directions of the wave vector and Poynting's vector are $\hat{e}_3$ and $-\hat{e}_3'$, respectively. Thus k_t^{I} must be in the second quadrant so that the Poynting's vector is pointing away from the interface, as shown in Fig. 4.3. Negative refraction occurs because of the negative refraction angle of the energy flow.

5 Negative Refraction in Bianisotropic Metamaterials

5.1 Introduction

Bianisotropic metamaterials have the most complicated constitutive parameters, where each of the four constitutive parameter tensors has nine elements, respectively. For simplicity we study the metamaterials where only the diagonal elements in the constitutive parameter tensors are nonzero. Specifically, if the three diagonal elements in each of the four constitutive parameters are equal, this kind of bianisotropic media reduces to a kind of biisotropic chiral media by letting $\xi \rightarrow i\xi$ and $\zeta \rightarrow -i\xi$. In this section, we will introduce the basic theoretical concepts and design principles of negative refraction in biisotropic chiral metamaterials.

5.2 Waves in Biisotropic Chiral Media

Biisotropic chiral media are a special kind of bianisotropic media. The source-free Maxwell equations in the biisotropic chiral media in the xyz coordinate system are

$$\bar{k} \times \bar{E} = \omega \bar{B}, \tag{5.1a}$$
$$\bar{k} \times \bar{H} = -\omega \bar{D}, \tag{5.1b}$$
$$\bar{k} \cdot \bar{D} = 0, \tag{5.1c}$$
$$\bar{k} \cdot \bar{B} = 0, \tag{5.1d}$$

and the constitutive relations are

$$\bar{D} = \varepsilon \bar{E} + i\xi \bar{H}, \tag{5.2a}$$
$$\bar{B} = -i\xi \bar{E} + \mu \bar{H}, \tag{5.2b}$$

in the $\overline{EH}$ representation, where ε is the permittivity, μ is the permeability and ξ is the chirality parameter. In the following we will derive the dispersion relations of waves in the biisotropic chiral media in the *kDB* coordinate system.

First, we need to transform the constitutive relations from the $\overline{EH}$ representation in the *xyz* coordinate system to the $\overline{DB}$ representation in the *xyz* coordinate system. According to Eqs. (2.25a, 2.25b, 2.26a, 2.26b, 2.26c, 2.26d, 2.27a, 2.27b, 2.27c, 2.27d), the constitutive relations in the $\overline{DB}$ representation in the *xyz* coordinate system are

$$\bar{E} = \kappa \bar{D} + i\chi \bar{B}, \tag{5.3a}$$
$$\bar{H} = -i\chi \bar{D} + \nu \bar{B}, \tag{5.3b}$$

where

$$\kappa = \left(\xi^{-1}\varepsilon - \mu^{-1}\xi\right)^{-1} \xi^{-1}, \tag{5.4a}$$
$$\chi = \left(\mu^{-1}\xi - \xi^{-1}\varepsilon\right)^{-1} \mu^{-1}, \tag{5.4b}$$
$$\nu = \left(\xi^{-1}\mu - \varepsilon^{-1}\xi\right)^{-1} \xi^{-1}, \tag{5.4c}$$

and

$$\varepsilon = \left(\chi^{-1}\kappa - \nu^{-1}\chi\right)^{-1} \chi^{-1}, \tag{5.5a}$$
$$\xi = \left(\nu^{-1}\chi - \chi^{-1}\kappa\right)^{-1} \nu^{-1}, \tag{5.5b}$$
$$\mu = \left(\chi^{-1}\nu - \kappa^{-1}\chi\right)^{-1} \chi^{-1}. \tag{5.5c}$$

Second, we need to transform the Maxwell equations and constitutive relations from the *xyz* coordinate system to the *kDB* coordinate system. From Eqs. (2.32a, 2.32b, 2.32c, 2.32d), the Maxwell equations in the biisotropic chiral media in the *kDB* coordinate system are

$$\bar{k}_k \times \bar{E}_k = \omega \bar{B}_k, \tag{5.6a}$$
$$\bar{k}_k \times \bar{H}_k = -\omega \bar{D}_k, \tag{5.6b}$$

$$\overline{k}_k \cdot \overline{D}_k = 0, \tag{5.6c}$$

$$\overline{k}_k \cdot \overline{B}_k = 0, \tag{5.6d}$$

and the constitutive relations are

$$\overline{E}_k = \kappa_k \overline{D}_k + i\chi_k \overline{B}_k, \tag{5.7a}$$

$$\overline{H}_k = -i\chi_k \overline{D}_k + \nu_k \overline{B}_k, \tag{5.7b}$$

where

$$\kappa_k = \kappa, \tag{5.8a}$$

$$\chi_k = \chi, \tag{5.8b}$$

$$\nu_k = \nu, \tag{5.8c}$$

according to Eqs. (2.28a, 2.28b, 2.29a, 2.29b, 2.29c, 2.29d).

Third, we need to derive the dispersion relations from the coefficient matrix. According to Eqs. (2.35, 2.36a, 2.36b, 2.36c, 2.36d, 5.8a, 5.8b, 5.8c), the determinant of the coefficient matrix is

$$\left|\overline{\overline{\kappa}}'_k - \left(\overline{\overline{i\chi}}'_k - iu\overline{\overline{\sigma}}_y\right) \overline{\overline{\nu}}'^{-1}_k \left(-\overline{\overline{i\chi}}'_k + iu\overline{\overline{\sigma}}_y\right)\right| = 0, \tag{5.9}$$

where

$$\overline{\overline{\kappa}}'_k = \begin{pmatrix} \kappa_k & 0 \\ 0 & \kappa_k \end{pmatrix} = \begin{pmatrix} \kappa & 0 \\ 0 & \kappa \end{pmatrix}, \tag{5.10a}$$

$$\overline{\overline{\chi}}'_k = \begin{pmatrix} \chi_k & 0 \\ 0 & \chi_k \end{pmatrix} = \begin{pmatrix} \chi & 0 \\ 0 & \chi \end{pmatrix}, \tag{5.10b}$$

$$\overline{\overline{\nu}}'_k = \begin{pmatrix} \nu_k & 0 \\ 0 & \nu_k \end{pmatrix} = \begin{pmatrix} \nu & 0 \\ 0 & \nu \end{pmatrix}, \tag{5.10c}$$

$$\overline{\overline{\sigma}}_y = \begin{pmatrix} 0 & -i \\ i & 0 \end{pmatrix}, \tag{5.10d}$$

and $u = \omega/k$. Thus the dispersion relations are

$$u^{\mathrm{I}} = \sqrt{\kappa\nu} + \chi, \tag{5.11a}$$

$$u^{\mathrm{II}} = \sqrt{\kappa\nu} - \chi. \tag{5.11b}$$

According to Eqs. (2.34a, 2.34b), the two dispersion relations correspond to two types of characteristic waves with $D_1/D_2 = -i$ and $D_1/D_2 = i$, respectively. We define the waves with $D_1/D_2 = -i$ as the *type-I waves*, which are right-handed circularly polarized waves, and the waves with $D_1/D_2 = i$ as the *type-II waves*, which are left-handed circularly polarized waves. It is worth noting that in this section we let $\kappa\nu > 0$ for both the type-I and type-II waves in order to study the propagating waves. Meanwhile, we only consider the case for $\chi > 0$. The case for $\chi < 0$ can be discussed by following a similar approach.

Fourth, we study the type-I and type-II waves in the *kDB* coordinate system, respectively. For a type-I wave, the wave vector is

$$\overline{k}_k^{\mathrm{I}} = \hat{e}_3 k^{\mathrm{I}} = \hat{e}_3 \frac{\omega}{\sqrt{\kappa\nu} + \chi}, \tag{5.12}$$

which is always positive since $\chi > 0$. From $D_1/D_2 = -i$, the electric flux density is

$$\overline{D}_k^{\mathrm{I}} = (\hat{e}_1 + i\hat{e}_2)\, D^{\mathrm{I}}. \tag{5.13a}$$

According to Eq. (2.34a) or Eq. (2.34b), the magnetic flux density is

$$\overline{B}_k^{\mathrm{I}} = \begin{cases} (-i\hat{e}_1 + \hat{e}_2)\sqrt{\kappa/\nu} D^{\mathrm{I}}, & \kappa > 0 \text{ and } \nu > 0, \\ (i\hat{e}_1 - \hat{e}_2)\sqrt{\kappa/\nu} D^{\mathrm{I}}, & \kappa < 0 \text{ and } \nu < 0. \end{cases} \tag{5.13b}$$

According to Eq. (5.7a), the electric field is

$$\overline{E}_k^{\mathrm{I}} = \begin{cases} (\hat{e}_1 + i\hat{e}_2)\,\kappa_+ D^{\mathrm{I}}, & \kappa > 0 \text{ and } \nu > 0, \\ (\hat{e}_1 + i\hat{e}_2)\,\kappa_- D^{\mathrm{I}}, & \kappa < 0 \text{ and } \nu < 0, \end{cases} \tag{5.13c}$$

where $\kappa_\pm = \kappa \pm \chi\sqrt{\kappa/\nu}$. Then, according to Eq. (5.7b), the magnetic field is

$$\overline{H}_k^{\mathrm{I}} = (-i\hat{e}_1 + \hat{e}_2)\left(\sqrt{\kappa\nu} + \chi\right) D^{\mathrm{I}}. \tag{5.13d}$$

Similarly, for a type-II wave, the wave vector is

$$\overline{k}_k^{\mathrm{II}} = \hat{e}_3 k^{\mathrm{II}} = \hat{e}_3 \frac{\omega}{\sqrt{\kappa\nu} - \chi}, \tag{5.14}$$

where $k^{\mathrm{II}} > 0$ for $\kappa\nu > \chi^2$ and $k^{\mathrm{II}} < 0$ for $\kappa\nu < \chi^2$. The field components are

$$\overline{D}_k^{\mathrm{II}} = (i\hat{e}_1 + \hat{e}_2)\, D^{\mathrm{II}}, \tag{5.15a}$$

$$\overline{B}_k^{\mathrm{II}} = \begin{cases} (-\hat{e}_1 + i\hat{e}_2)\sqrt{\kappa/\nu} D^{\mathrm{II}}, & \kappa > 0 \text{ and } \nu > 0, \\ (\hat{e}_1 - i\hat{e}_2)\sqrt{\kappa/\nu} D^{\mathrm{II}}, & \kappa < 0 \text{ and } \nu < 0, \end{cases} \tag{5.15b}$$

$$\overline{E}_k^{\mathrm{II}} = \begin{cases} (i\hat{e}_1 + \hat{e}_2)\,\kappa_- D^{\mathrm{II}}, & \kappa > 0 \text{ and } \nu > 0, \\ (i\hat{e}_1 + \hat{e}_2)\,\kappa_+ D^{\mathrm{II}}, & \kappa < 0 \text{ and } \nu < 0, \end{cases} \tag{5.15c}$$

$$\overline{H}_k^{\mathrm{II}} = (-\hat{e}_1 + i\hat{e}_2)\left(\sqrt{\kappa\nu} - \chi\right) D^{\mathrm{II}}. \tag{5.15d}$$

The Poynting's vectors in the *kDB* coordinate system for both the type-I and type-II waves can be calculated by

$$\langle \overline{S}_k \rangle = \frac{1}{2}\mathrm{Re}\left(\overline{E}_k \times \overline{H}_k^*\right). \tag{5.16}$$

From Eqs. (5.13c, 5.13d), for a type-I wave, we get

$$\left\langle \overline{S}_k^{\mathrm{I}} \right\rangle = \begin{cases} \hat{e}_3 \sqrt{\kappa/\nu} \left(\sqrt{\kappa\nu} + \chi\right)^2 |D^{\mathrm{I}}|^2, & \kappa > 0 \text{ and } \nu > 0, \\ -\hat{e}_3 \sqrt{\kappa/\nu} \left(\sqrt{\kappa\nu} + \chi\right)^2 |D^{\mathrm{I}}|^2, & \kappa < 0 \text{ and } \nu < 0. \end{cases} \tag{5.17}$$

Similarly, from Eqs. (5.15c, 5.15d), for a type-II wave, we get

$$\left\langle \overline{S}_k^{\mathrm{II}} \right\rangle = \begin{cases} \hat{e}_3 \sqrt{\kappa/\nu} \left(\sqrt{\kappa\nu} - \chi\right)^2 |D^{\mathrm{II}}|^2, & \kappa > 0 \text{ and } \nu > 0, \\ -\hat{e}_3 \sqrt{\kappa/\nu} \left(\sqrt{\kappa\nu} - \chi\right)^2 |D^{\mathrm{II}}|^2, & \kappa < 0 \text{ and } \nu < 0. \end{cases} \tag{5.18}$$

Thus, for any type of waves, there are two sets of field components, which correspond to $\kappa > 0$, $\nu > 0$ and $\kappa < 0$, $\nu < 0$, respectively. In the following we define the media with $\kappa > 0$ and $\nu > 0$ as the *conventional materials*, and the media with $\kappa < 0$ and $\nu < 0$ as the *metamaterials*. The type-I and type-II waves constitute a set of eigenstates in biisotropic chiral media, and any waves can be expanded using this set of eigenstates.

Finally, we transform the field components from the *kDB* coordinate system back to the *xyz* coordinate system. The coordinate systems are shown in Fig. 2.1. According to Eqs. (2.22a, 2.22b, 2.22c, 2.22d), for the type-I waves, we have

$$\begin{aligned} \overline{D}^{\mathrm{I}} &= \hat{x}\left(\sin\phi + i\cos\theta\cos\phi\right) D^{\mathrm{I}} \\ &\quad + \hat{y}\left(-\cos\phi + i\cos\theta\sin\phi\right) D^{\mathrm{I}} - \hat{z} i \sin\theta D^{\mathrm{I}}, \end{aligned} \tag{5.19a}$$

$$\begin{aligned} \overline{E}^{\mathrm{I}} &= \hat{x}\left(\sin\phi + i\cos\theta\cos\phi\right) \kappa_+ D^{\mathrm{I}} \\ &\quad + \hat{y}\left(-\cos\phi + i\cos\theta\sin\phi\right) \kappa_+ D^{\mathrm{I}} - \hat{z} i \sin\theta \kappa_+ D^{\mathrm{I}}, \end{aligned} \tag{5.19b}$$

$$\begin{aligned} \overline{B}^{\mathrm{I}} &= \hat{x}\left(-i\sin\phi + \cos\theta\cos\phi\right) \sqrt{\kappa/\nu} D^{\mathrm{I}} \\ &\quad + \hat{y}\left(i\cos\phi + \cos\theta\sin\phi\right) \sqrt{\kappa/\nu} D^{\mathrm{I}} - \hat{z}\sin\theta \sqrt{\kappa/\nu} D^{\mathrm{I}}, \end{aligned} \tag{5.19c}$$

$$\begin{aligned} \overline{H}^{\mathrm{I}} &= \hat{x}\left(-i\sin\phi + \cos\theta\cos\phi\right) \left(\sqrt{\kappa\nu} + \chi\right) D^{\mathrm{I}} \\ &\quad + \hat{y}\left(i\cos\phi + \cos\theta\sin\phi\right) \left(\sqrt{\kappa\nu} + \chi\right) D^{\mathrm{I}} - \hat{z}\sin\theta \left(\sqrt{\kappa\nu} + \chi\right) D^{\mathrm{I}}, \end{aligned} \tag{5.19d}$$

in the conventional materials, and

$$\begin{aligned} \overline{D}^{\mathrm{I}} &= \hat{x}\left(\sin\phi + i\cos\theta\cos\phi\right) D^{\mathrm{I}} \\ &\quad + \hat{y}\left(-\cos\phi + i\cos\theta\sin\phi\right) D^{\mathrm{I}} - \hat{z} i \sin\theta D^{\mathrm{I}}, \end{aligned} \tag{5.20a}$$

$$\begin{aligned} \overline{E}^{\mathrm{I}} &= \hat{x}\left(\sin\phi + i\cos\theta\cos\phi\right) \kappa_- D^{\mathrm{I}} \\ &\quad + \hat{y}\left(-\cos\phi + i\cos\theta\sin\phi\right) \kappa_- D^{\mathrm{I}} - \hat{z} i \sin\theta \kappa_- D^{\mathrm{I}}, \end{aligned} \tag{5.20b}$$

$$\overline{B}^{\mathrm{I}} = \hat{x}\,(i\sin\phi - \cos\theta\cos\phi)\,\sqrt{\kappa/\nu}D^{\mathrm{I}} + \hat{y}\,(-i\cos\phi - \cos\theta\sin\phi)\,\sqrt{\kappa/\nu}D^{\mathrm{I}} + \hat{z}\sin\theta\,\sqrt{\kappa/\nu}D^{\mathrm{I}}, \tag{5.20c}$$

$$\overline{H}^{\mathrm{I}} = \hat{x}\,(-i\sin\phi + \cos\theta\cos\phi)\left(\sqrt{\kappa\nu} + \chi\right)D^{\mathrm{I}} + \hat{y}\,(i\cos\phi + \cos\theta\sin\phi)\left(\sqrt{\kappa\nu} + \chi\right)D^{\mathrm{I}} - \hat{z}\sin\theta\left(\sqrt{\kappa\nu} + \chi\right)D^{\mathrm{I}}, \tag{5.20d}$$

in the metamaterials, respectively. Similarly, for the type-II waves, we have

$$\overline{D}^{\mathrm{II}} = \hat{x}\,(i\sin\phi + \cos\theta\cos\phi)\,D^{\mathrm{II}} + \hat{y}\,(-i\cos\phi + \cos\theta\sin\phi)\,D^{\mathrm{II}} - \hat{z}\sin\theta D^{\mathrm{II}}, \tag{5.21a}$$

$$\overline{E}^{\mathrm{II}} = \hat{x}\,(i\sin\phi + \cos\theta\cos\phi)\,\kappa_{-}D^{\mathrm{II}} + \hat{y}\,(-i\cos\phi + \cos\theta\sin\phi)\,\kappa_{-}D^{\mathrm{II}} - \hat{z}\sin\theta\kappa_{-}D^{\mathrm{II}}, \tag{5.21b}$$

$$\overline{B}^{\mathrm{II}} = \hat{x}\,(-\sin\phi + i\cos\theta\cos\phi)\,\sqrt{\kappa/\nu}D^{\mathrm{II}} + \hat{y}\,(\cos\phi + i\cos\theta\sin\phi)\,\sqrt{\kappa/\nu}D^{\mathrm{II}} - \hat{z}i\sin\theta\,\sqrt{\kappa/\nu}D^{\mathrm{II}}, \tag{5.21c}$$

$$\overline{H}^{\mathrm{II}} = \hat{x}\,(-\sin\phi + i\cos\theta\cos\phi)\left(\sqrt{\kappa\nu} - \chi\right)D^{\mathrm{II}} + \hat{y}\,(\cos\phi + i\cos\theta\sin\phi)\left(\sqrt{\kappa\nu} - \chi\right)D^{\mathrm{II}} - \hat{z}i\sin\theta\left(\sqrt{\kappa\nu} - \chi\right)D^{\mathrm{II}}, \tag{5.21d}$$

in the conventional materials, and

$$\overline{D}^{\mathrm{II}} = \hat{x}\,(i\sin\phi + \cos\theta\cos\phi)\,D^{\mathrm{II}} + \hat{y}\,(-i\cos\phi + \cos\theta\sin\phi)\,D^{\mathrm{II}} - \hat{z}\sin\theta D^{\mathrm{II}}, \tag{5.22a}$$

$$\overline{E}^{\mathrm{II}} = \hat{x}\,(i\sin\phi + \cos\theta\cos\phi)\,\kappa_{+}D^{\mathrm{II}} + \hat{y}\,(-i\cos\phi + \cos\theta\sin\phi)\,\kappa_{+}D^{\mathrm{II}} - \hat{z}\sin\theta\kappa_{+}D^{\mathrm{II}}, \tag{5.22b}$$

$$\overline{B}^{\mathrm{II}} = \hat{x}\,(\sin\phi - i\cos\theta\cos\phi)\,\sqrt{\kappa/\nu}D^{\mathrm{II}} - \hat{y}\,(\cos\phi + i\cos\theta\sin\phi)\,\sqrt{\kappa/\nu}D^{\mathrm{II}} + \hat{z}i\sin\theta\,\sqrt{\kappa/\nu}D^{\mathrm{II}}, \tag{5.22c}$$

$$\overline{H}^{\mathrm{II}} = \hat{x}\,(-\sin\phi + i\cos\theta\cos\phi)\left(\sqrt{\kappa\nu} - \chi\right)D^{\mathrm{II}} + \hat{y}\,(\cos\phi + i\cos\theta\sin\phi)\left(\sqrt{\kappa\nu} - \chi\right)D^{\mathrm{II}} - \hat{z}i\sin\theta\left(\sqrt{\kappa\nu} - \chi\right)D^{\mathrm{II}}, \tag{5.22d}$$

in the metamaterials, respectively. Note that the above field components are expressed in the $\overline{DB}$ representation.

Transforming from the $\overline{DB}$ representation to the $\overline{EH}$ representation, the dispersion relations are

$$k^{\mathrm{I}} = \omega\left(\sqrt{\varepsilon\mu} + \xi\right), \tag{5.23a}$$

$$k^{\mathrm{II}} = \omega\left(\sqrt{\varepsilon\mu} - \xi\right), \tag{5.23b}$$

for $\varepsilon\mu > \xi^2$, and

$$k^{\mathrm{I}} = -\omega\left(\sqrt{\varepsilon\mu} - \xi\right), \tag{5.24a}$$

$$k^{\mathrm{II}} = -\omega\left(\sqrt{\varepsilon\mu} + \xi\right) \tag{5.24b}$$

for $\varepsilon\mu < \xi^2$, respectively, according to Eqs. (5.4a, 5.4b, 5.4c, 5.11a, 5.11b). Meanwhile, from $\chi > 0$ we get $\xi < 0$ for $\varepsilon\mu > \xi^2$ and $\xi > 0$ for $\varepsilon\mu < \xi^2$. We get four cases in total. For case 1 the conventional materials with $\kappa > 0$ and $\nu > 0$ require that $\varepsilon\mu > \xi^2$ with $\varepsilon > 0$ and $\mu > 0$. For case 2 the metamaterials with $\kappa < 0$ and $\nu < 0$ require that $\varepsilon\mu < \xi^2$ with $\varepsilon > 0$ and $\mu > 0$. For case 3 the conventional materials with $\kappa > 0$ and $\nu > 0$ require that $\varepsilon\mu < \xi^2$ with $\varepsilon < 0$ and $\mu < 0$. For case 4 the metamaterials with $\kappa < 0$ and $\nu < 0$ require that $\varepsilon\mu > \xi^2$ with $\varepsilon < 0$ and $\mu < 0$. For a type-I wave, the wave number is

$$k^{\mathrm{I}} = \begin{cases} \omega\left(\sqrt{\varepsilon\mu} + \xi\right), & \varepsilon\mu > \xi^2, \\ -\omega\left(\sqrt{\varepsilon\mu} - \xi\right), & \varepsilon\mu < \xi^2, \end{cases} \tag{5.25}$$

which is always positive. For brevity the field components are omitted here. The readers can easily derive them by themselves. Since $\bar{k} = \hat{e}_3 k = \hat{x}k\sin\theta\cos\phi + \hat{y}k\sin\theta\sin\phi + \hat{z}k\cos\theta$ and $\bar{r} = \hat{x}x + \hat{y}y + \hat{z}z$, the field components contain a phase factor, where

$$D^{\mathrm{I}} = D_0^{\mathrm{I}} e^{i\bar{k}^{\mathrm{I}}\cdot\bar{r}} = D_0^{\mathrm{I}} e^{ik^{\mathrm{I}}(x\sin\theta\cos\phi + y\sin\theta\sin\phi + z\cos\theta)}. \tag{5.26}$$

Similarly, for a type-II wave, the wave number is

$$k^{\mathrm{II}} = \begin{cases} \omega\left(\sqrt{\varepsilon\mu} - \xi\right), & \varepsilon\mu > \xi^2, \\ -\omega\left(\sqrt{\varepsilon\mu} + \xi\right), & \varepsilon\mu < \xi^2. \end{cases} \tag{5.27}$$

Considering $\xi < 0$ for $\varepsilon\mu > \xi^2$ and $\xi > 0$ for $\varepsilon\mu < \xi^2$, we have $k^{\mathrm{II}} > 0$ for $\varepsilon\mu > \xi^2$ and $k^{\mathrm{II}} < 0$ for $\varepsilon\mu < \xi^2$. The field components are also omitted here, and they contain a phase factor, where

$$D^{\mathrm{II}} = D_0^{\mathrm{II}} e^{i\bar{k}^{\mathrm{II}}\cdot\bar{r}} = D_0^{\mathrm{II}} e^{ik^{\mathrm{II}}(x\sin\theta\cos\phi + y\sin\theta\sin\phi + z\cos\theta)}. \tag{5.28}$$

The Poynting's vectors can be calculated by

$$\langle\bar{S}\rangle = \frac{1}{2}\mathrm{Re}\left(\bar{E}\times\bar{H}^*\right). \tag{5.29}$$

For the type-I waves, we get

$$\left\langle\bar{S}^{\mathrm{I}}\right\rangle = \begin{cases} \hat{e}_3\dfrac{\eta}{\left(\sqrt{\varepsilon\mu}+\xi\right)^2}D_0^{\mathrm{I}2}, & \varepsilon > 0, \mu > 0 \text{ and } \varepsilon\mu > \xi^2, \\ \hat{e}_3\dfrac{\eta}{\left(\sqrt{\varepsilon\mu}-\xi\right)^2}D_0^{\mathrm{I}2}, & \varepsilon < 0, \mu < 0 \text{ and } \varepsilon\mu < \xi^2, \\ -\hat{e}_3\dfrac{\eta}{\left(\sqrt{\varepsilon\mu}-\xi\right)^2}D_0^{\mathrm{I}2}, & \varepsilon > 0, \mu > 0 \text{ and } \varepsilon\mu < \xi^2, \\ -\hat{e}_3\dfrac{\eta}{\left(\sqrt{\varepsilon\mu}+\xi\right)^2}D_0^{\mathrm{I}2}, & \varepsilon < 0, \mu < 0 \text{ and } \varepsilon\mu > \xi^2, \end{cases} \tag{5.30}$$

Table 5.1 The main properties of type-I waves in biisotropic chiral media.

type I wave	direction of $\overline{\mathbf{k}}^{\mathrm{I}}$	direction of $\langle \overline{\mathbf{S}}^{\mathrm{I}} \rangle$
case 1: $\varepsilon > 0$, $\mu > 0$ and $\varepsilon\mu > \xi^2$	$\hat{e}_3$	$\hat{e}_3$
case 2: $\varepsilon > 0$, $\mu > 0$ and $\varepsilon\mu < \xi^2$	$\hat{e}_3$	$-\hat{e}_3$
case 3: $\varepsilon < 0$, $\mu < 0$ and $\varepsilon\mu > \xi^2$	$\hat{e}_3$	$-\hat{e}_3$
case 4: $\varepsilon < 0$, $\mu < 0$ and $\varepsilon\mu < \xi^2$	$\hat{e}_3$	$\hat{e}_3$

Table 5.2 The main properties of type-II waves in biisotropic chiral media.

type II wave	direction of $\overline{\mathbf{k}}^{\mathrm{II}}$	direction of $\langle \overline{\mathbf{S}}^{\mathrm{II}} \rangle$
case 1: $\varepsilon > 0$, $\mu > 0$ and $\varepsilon\mu > \xi^2$	$\hat{e}_3$	$\hat{e}_3$
case 2: $\varepsilon > 0$, $\mu > 0$ and $\varepsilon\mu < \xi^2$	$-\hat{e}_3$	$-\hat{e}_3$
case 3: $\varepsilon < 0$, $\mu < 0$ and $\varepsilon\mu > \xi^2$	$\hat{e}_3$	$-\hat{e}_3$
case 4: $\varepsilon < 0$, $\mu < 0$ and $\varepsilon\mu < \xi^2$	$-\hat{e}_3$	$\hat{e}_3$

where $\hat{e}_3 = \hat{x}\sin\theta\cos\phi + \hat{y}\sin\theta\sin\phi + \hat{z}\cos\theta$. Similarly, for the type-II waves, we get

$$\left\langle \overline{S}^{\mathrm{II}} \right\rangle = \begin{cases} \hat{e}_3 \frac{\eta}{\left(\sqrt{\varepsilon\mu}-\xi\right)^2} D_0^{\mathrm{II}2}, & \varepsilon > 0, \mu > 0 \text{ and } \varepsilon\mu > \xi^2, \\ \hat{e}_3 \frac{\eta}{\left(\sqrt{\varepsilon\mu}+\xi\right)^2} D_0^{\mathrm{II}2}, & \varepsilon < 0, \mu < 0 \text{ and } \varepsilon\mu < \xi^2, \\ -\hat{e}_3 \frac{\eta}{\left(\sqrt{\varepsilon\mu}+\xi\right)^2} D_0^{\mathrm{II}2}, & \varepsilon > 0, \mu > 0 \text{ and } \varepsilon\mu < \xi^2, \\ -\hat{e}_3 \frac{\eta}{\left(\sqrt{\varepsilon\mu}-\xi\right)^2} D_0^{\mathrm{II}2}, & \varepsilon < 0, \mu < 0 \text{ and } \varepsilon\mu > \xi^2, \end{cases} \tag{5.31}$$

where $\hat{e}_3 = \hat{x}\sin\theta\cos\phi + \hat{y}\sin\theta\sin\phi + \hat{z}\cos\theta$. In light of the directions of the wave vectors and Poynting's vectors, the main properties of type-I and type-II waves are summarized in Tabs. 5.1 and 5.2.

5.3 Negative Refraction in Biisotropic Chiral Metamaterials

In this section we show that negative refraction can occur at an interface between two biisotropic chiral media. We let the incident plane be the *xz* plane and the boundary between the two media be the *yz* plane, as shown in Fig. 5.1. All the angles are assumed to have positive values.

As an example we only consider the incidence of a type-I wave and its refraction from a medium in case 1 with $\varepsilon > 0$, $\mu > 0$ and $\varepsilon\mu > \xi^2$ to a medium in case 3 with $\varepsilon < 0$, $\mu < 0$ and $\varepsilon\mu > \xi^2$. The field components and Poynting's vectors for both the type-I and type-II waves in the two media were derived at

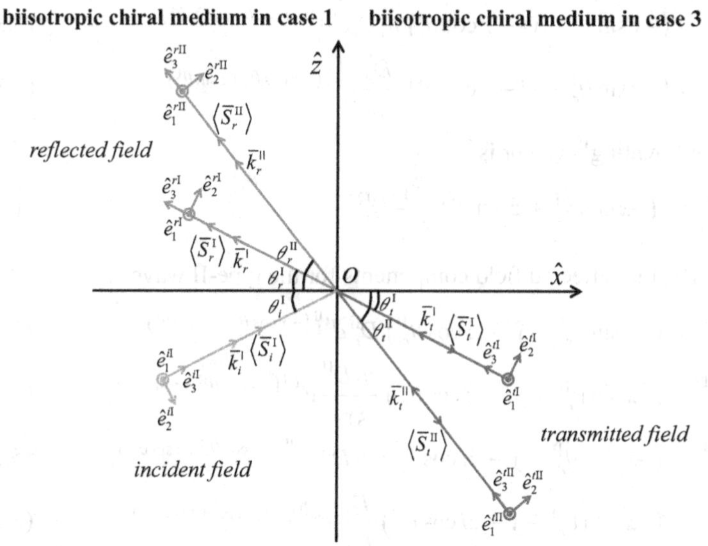

Figure 5.1 A schematic of the negative refraction at an interface between a biisotropic chiral medium in case 1 and a biisotropic chiral medium in case 3.

the end of the last section. For the incident field, $\theta = \pi/2 - \theta_i^{\mathrm{I}}$ and $\phi = 0$. Then the incident field components are

$$\overline{D}_i^{\mathrm{I}} = \left(\hat{x}i \sin\theta_i^{\mathrm{I}} - \hat{y} - \hat{z}i\cos\theta_i^{\mathrm{I}}\right) D_i^{\mathrm{I}} e^{ik_1^{\mathrm{I}}\left(x\cos\theta_i^{\mathrm{I}} + z\sin\theta_i^{\mathrm{I}}\right)}, \tag{5.32a}$$

$$\overline{E}_i^{\mathrm{I}} = \left(\hat{x}i \sin\theta_i^{\mathrm{I}} - \hat{y} - \hat{z}i\cos\theta_i^{\mathrm{I}}\right) \frac{\eta_1 D_i^{\mathrm{I}}}{\xi_{1+}} e^{ik_1^{\mathrm{I}}\left(x\cos\theta_i^{\mathrm{I}} + z\sin\theta_i^{\mathrm{I}}\right)}, \tag{5.32b}$$

$$\overline{B}_i^{\mathrm{I}} = \left(\hat{x} \sin\theta_i^{\mathrm{I}} + \hat{y}i - \hat{z}\cos\theta_i^{\mathrm{I}}\right) \eta_1 D_i^{\mathrm{I}} e^{ik_1^{\mathrm{I}}\left(x\cos\theta_i^{\mathrm{I}} + z\sin\theta_i^{\mathrm{I}}\right)}, \tag{5.32c}$$

$$\overline{H}_i^{\mathrm{I}} = \left(\hat{x} \sin\theta_i^{\mathrm{I}} + \hat{y}i - \hat{z}\cos\theta_i^{\mathrm{I}}\right) \frac{D_i^{\mathrm{I}}}{\xi_{1+}} e^{ik_1^{\mathrm{I}}\left(x\cos\theta_i^{\mathrm{I}} + z\sin\theta_i^{\mathrm{I}}\right)}, \tag{5.32d}$$

and the Poynting's vector is

$$\left\langle \overline{S}_i^{\mathrm{I}} \right\rangle = \left(\hat{x}\cos\theta_i^{\mathrm{I}} + \hat{z}\sin\theta_i^{\mathrm{I}}\right) \frac{\eta_1}{\xi_{1+}^2} D_i^{\mathrm{I}2}, \tag{5.33}$$

where $\eta_1 = \sqrt{\mu_1/\varepsilon_1}$ is the impedance, $k_1^{\mathrm{I}} = \omega\left(\sqrt{\varepsilon_1\mu_1} + \xi_1\right)$ is positive and $\xi_{1+} = \sqrt{\varepsilon_1\mu_1} + \xi_1$. For the reflected fields, $\theta = \pi/2 - \theta_r^{\mathrm{I}}$ and $\phi = \pi$ for the type-I wave and $\theta = \pi/2 - \theta_r^{\mathrm{II}}$ and $\phi = \pi$ for the type-II wave. Then the reflected field components for the type-I wave are

$$\overline{D}_r^{\mathrm{I}} = \left(-\hat{x}i \sin\theta_r^{\mathrm{I}} + \hat{y} - \hat{z}i\cos\theta_r^{\mathrm{I}}\right) D_r^{\mathrm{I}} e^{ik_1^{\mathrm{I}}\left(-x\cos\theta_r^{\mathrm{I}} + z\sin\theta_r^{\mathrm{I}}\right)}, \tag{5.34a}$$

$$\overline{E}_r^{\mathrm{I}} = \left(-\hat{x}i \sin\theta_r^{\mathrm{I}} + \hat{y} - \hat{z}i\cos\theta_r^{\mathrm{I}}\right) \frac{\eta_1 D_r^{\mathrm{I}}}{\xi_{1+}} e^{ik_1^{\mathrm{I}}\left(-x\cos\theta_r^{\mathrm{I}} + z\sin\theta_r^{\mathrm{I}}\right)}, \tag{5.34b}$$

$$\overline{B}_r^{\rm I} = \left(-\hat{x}\sin\theta_r^{\rm I} - \hat{y}i - \hat{z}\cos\theta_r^{\rm I}\right)\eta_1 D_r^{\rm I} e^{ik_1^{\rm I}\left(-x\cos\theta_r^{\rm I}+z\sin\theta_r^{\rm I}\right)}, \tag{5.34c}$$

$$\overline{H}_r^{\rm I} = \left(-\hat{x}\sin\theta_r^{\rm I} - \hat{y}i - \hat{z}\cos\theta_r^{\rm I}\right)\frac{D_r^{\rm I}}{\xi_{1+}} e^{ik_1^{\rm I}\left(-x\cos\theta_r^{\rm I}+z\sin\theta_r^{\rm I}\right)}, \tag{5.34d}$$

and the Poynting's vector is

$$\left\langle \overline{S}_r^{\rm I}\right\rangle = \left(-\hat{x}\cos\theta_r^{\rm I} + \hat{z}\sin\theta_r^{\rm I}\right)\frac{\eta_1}{\xi_{1+}^2} D_r^{\rm I2}. \tag{5.35}$$

Similarly, the reflected field components for the type-II wave are

$$\overline{D}_r^{\rm II} = \left(-\hat{x}\sin\theta_r^{\rm II} + \hat{y}i - \hat{z}\cos\theta_r^{\rm II}\right) D_r^{\rm II} e^{ik_1^{\rm II}\left(-x\cos\theta_r^{\rm II}+z\sin\theta_r^{\rm II}\right)}, \tag{5.36a}$$

$$\overline{E}_r^{\rm II} = \left(-\hat{x}\sin\theta_r^{\rm II} + \hat{y}i - \hat{z}\cos\theta_r^{\rm II}\right)\frac{\eta_1 D_r^{\rm II}}{\xi_{1-}} e^{ik_1^{\rm II}\left(-x\cos\theta_r^{\rm II}+z\sin\theta_r^{\rm II}\right)}, \tag{5.36b}$$

$$\overline{B}_r^{\rm II} = \left(-\hat{x}i\sin\theta_r^{\rm II} - \hat{y} - \hat{z}i\cos\theta_r^{\rm II}\right)\eta_1 D_r^{\rm II} e^{ik_1^{\rm II}\left(-x\cos\theta_r^{\rm II}+z\sin\theta_r^{\rm II}\right)}, \tag{5.36c}$$

$$\overline{H}_r^{\rm II} = \left(-\hat{x}i\sin\theta_r^{\rm II} - \hat{y} - \hat{z}i\cos\theta_r^{\rm II}\right)\frac{D_r^{\rm II}}{\xi_{1-}} e^{ik_1^{\rm II}\left(-x\cos\theta_r^{\rm II}+z\sin\theta_r^{\rm II}\right)}, \tag{5.36d}$$

and the Poynting's vector is

$$\left\langle \overline{S}_r^{\rm II}\right\rangle = \left(-\hat{x}\cos\theta_r^{\rm II} + \hat{z}\sin\theta_r^{\rm II}\right)\frac{\eta_1}{\xi_{1-}^2} D_r^{\rm II2}, \tag{5.37}$$

where $k_1^{\rm II} = \omega\left(\sqrt{\varepsilon_1\mu_1} - \xi_1\right)$ is positive and $\xi_{1-} = \sqrt{\varepsilon_1\mu_1} - \xi_1$. For the transmitted fields, $\theta = \pi/2 - \theta_t^{\rm I}$ and $\phi = \pi$ for the type-I wave and $\theta = \pi/2 - \theta_t^{\rm II}$ and $\phi = \pi$ for the type-II wave by considering the directions of the energy flows. Then the transmitted field components for the type-I wave are

$$\overline{D}_t^{\rm I} = \left(-\hat{x}i\sin\theta_t^{\rm I} + \hat{y} - \hat{z}i\cos\theta_t^{\rm I}\right) D_t^{\rm I} e^{ik_2^{\rm I}\left(-x\cos\theta_t^{\rm I}+z\sin\theta_t^{\rm I}\right)}, \tag{5.38a}$$

$$\overline{E}_t^{\rm I} = \left(\hat{x}i\sin\theta_t^{\rm I} - \hat{y} + \hat{z}i\cos\theta_t^{\rm I}\right)\frac{\eta_2 D_t^{\rm I}}{\xi_{2+}} e^{ik_2^{\rm I}\left(-x\cos\theta_t^{\rm I}+z\sin\theta_t^{\rm I}\right)}, \tag{5.38b}$$

$$\overline{B}_t^{\rm I} = \left(\hat{x}\sin\theta_t^{\rm I} + \hat{y}i + \hat{z}\cos\theta_t^{\rm I}\right)\eta_2 D_t^{\rm I} e^{ik_2^{\rm I}\left(-x\cos\theta_t^{\rm I}+z\sin\theta_t^{\rm I}\right)}, \tag{5.38c}$$

$$\overline{H}_t^{\rm I} = \left(-\hat{x}\sin\theta_t^{\rm I} - \hat{y}i - \hat{z}\cos\theta_t^{\rm I}\right)\frac{D_t^{\rm I}}{\xi_{2+}} e^{ik_2^{\rm I}\left(-x\cos\theta_t^{\rm I}+z\sin\theta_t^{\rm I}\right)}, \tag{5.38d}$$

and the Poynting's vector is

$$\left\langle \overline{S}_t^{\rm I}\right\rangle = \left(\hat{x}\cos\theta_t^{\rm I} - \hat{z}\sin\theta_t^{\rm I}\right)\frac{\eta_2}{\xi_{2+}^2} D_t^{\rm I2}, \tag{5.39}$$

where $\eta_2 = \sqrt{\mu_2/\varepsilon_2}$ is the impedance, $k_2^{\rm I} = \omega\left(\sqrt{\varepsilon_2\mu_2} + \xi_2\right)$ is positive and $\xi_{2+} = \sqrt{\varepsilon_2\mu_2} + \xi_2$. Similarly, the transmitted field components for the type-II wave are

$$\overline{D}_t^{\rm II} = \left(-\hat{x}\sin\theta_t^{\rm II} + \hat{y}i - \hat{z}\cos\theta_t^{\rm II}\right) D_t^{\rm II} e^{ik_2^{\rm II}\left(-x\cos\theta_t+z\sin\theta_t\right)}, \tag{5.40a}$$

$$\overline{E}_t^{\rm II} = \left(\hat{x}\sin\theta_t^{\rm II} - \hat{y}i + \hat{z}\cos\theta_t^{\rm II}\right)\frac{\eta_2 D_t^{\rm II}}{\xi_{2-}} e^{ik_2^{\rm II}\left(-x\cos\theta_t+z\sin\theta_t\right)}, \tag{5.40b}$$

$$\overline{B}_t^{\mathrm{II}} = \left(\hat{x}i\sin\theta_t^{\mathrm{II}} + \hat{y} + \hat{z}i\cos\theta_t^{\mathrm{II}}\right)\eta_2 D_t^{\mathrm{II}} e^{ik_2^{\mathrm{II}}(-x\cos\theta_t + z\sin\theta_t)}, \tag{5.40c}$$

$$\overline{H}_t^{\mathrm{II}} = \left(-\hat{x}i\sin\theta_t^{\mathrm{II}} - \hat{y} - \hat{z}i\cos\theta_t^{\mathrm{II}}\right)\frac{D_t^{\mathrm{II}}}{\xi_{2-}} e^{ik_2^{\mathrm{II}}(-x\cos\theta_t + z\sin\theta_t)}, \tag{5.40d}$$

and the Poynting's vector is

$$\left\langle \overline{S}_t^{\mathrm{II}} \right\rangle = \left(\hat{x}\cos\theta_t^{\mathrm{II}} - \hat{z}\sin\theta_t^{\mathrm{II}}\right)\frac{\eta_2}{\xi_{2-}^2} D_t^{\mathrm{II}2}, \tag{5.41}$$

where $k_2^{\mathrm{II}} = \omega\left(\sqrt{\varepsilon_2\mu_2} - \xi_2\right)$ is positive and $\xi_{2-} = \sqrt{\varepsilon_2\mu_2} - \xi_2$. Note that the direction of the energy flow is always from the interface to infinity.

By considering the continuity conditions of the electric fields and magnetic fields, we get

$$-\frac{\eta_1}{\xi_{1+}} D_i^{\mathrm{I}} + \frac{\eta_1}{\xi_{1+}} D_r^{\mathrm{I}} + i\frac{\eta_1}{\xi_{1-}} D_r^{\mathrm{II}} = -\frac{\eta_2}{\xi_{2+}} D_t^{\mathrm{I}} - i\frac{\eta_2}{\xi_{2-}} D_t^{\mathrm{II}}, \tag{5.42a}$$

$$-i\frac{\cos\theta_i^{\mathrm{I}}\eta_1}{\xi_{1+}} D_i^{\mathrm{I}} - i\frac{\cos\theta_r^{\mathrm{I}}\eta_1}{\xi_{1+}} D_r^{\mathrm{I}} - \frac{\cos\theta_r^{\mathrm{II}}\eta_1}{\xi_{1-}} D_r^{\mathrm{II}} = i\frac{\cos\theta_t^{\mathrm{I}}\eta_2}{\xi_{2+}} D_t^{\mathrm{I}} + \frac{\cos\theta_t^{\mathrm{II}}\eta_2}{\xi_{2-}} D_t^{\mathrm{II}}, \tag{5.42b}$$

$$i\frac{1}{\xi_{1+}} D_i^{\mathrm{I}} - i\frac{1}{\xi_{1+}} D_r^{\mathrm{I}} - \frac{1}{\xi_{1-}} D_r^{\mathrm{II}} = -i\frac{1}{\xi_{2+}} D_t^{\mathrm{I}} - \frac{1}{\xi_{2-}} D_t^{\mathrm{II}}, \tag{5.42c}$$

$$-\frac{\cos\theta_i^{\mathrm{I}}}{\xi_{1+}} D_i^{\mathrm{I}} - \frac{\cos\theta_r^{\mathrm{I}}}{\xi_{1+}} D_r^{\mathrm{I}} - i\frac{\cos\theta_r^{\mathrm{II}}}{\xi_{1-}} D_r^{\mathrm{II}} = -\frac{\cos\theta_t^{\mathrm{I}}}{\xi_{2-}} D_t^{\mathrm{I}} - i\frac{\cos\theta_t^{\mathrm{II}}}{\xi_{2-}} D_t^{\mathrm{II}}, \tag{5.42d}$$

and

$$k_1^{\mathrm{I}}\sin\theta_i^{\mathrm{I}} = k_1^{\mathrm{I}}\sin\theta_r^{\mathrm{I}} = k_1^{\mathrm{II}}\sin\theta_r^{\mathrm{II}} = k_2^{\mathrm{I}}\sin\theta_t^{\mathrm{I}} = k_2^{\mathrm{II}}\sin\theta_t^{\mathrm{II}}. \tag{5.43}$$

The latter equation is the continuity condition of the wave vectors. From Eqs. (5.42a, 5.42b, 5.42c, 5.42d), the solution for the coefficients is

$$\overline{D}_o = \overline{\overline{C}}^{-1}\overline{D}_i, \tag{5.44}$$

where

$$\overline{D}_o = \begin{pmatrix} D_r^{\mathrm{I}} \\ D_r^{\mathrm{II}} \\ D_t^{\mathrm{I}} \\ D_t^{\mathrm{II}} \end{pmatrix}, \tag{5.45a}$$

$$\overline{\overline{C}} = \begin{pmatrix} \frac{\eta_1}{\xi_{1+}} & \frac{i\eta_1}{\xi_{1-}} & \frac{\eta_2}{\xi_{2+}} & \frac{i\eta_2}{\xi_{2-}} \\ \frac{-i\cos\theta_r^{\mathrm{I}}\eta_1}{\xi_{1+}} & \frac{-\cos\theta_r^{\mathrm{II}}\eta_1}{\xi_{1-}} & \frac{-i\cos\theta_t^{\mathrm{I}}\eta_2}{\xi_{2+}} & \frac{-\cos\theta_t^{\mathrm{II}}\eta_2}{\xi_{2-}} \\ \frac{-i}{\xi_{1+}} & \frac{-1}{\xi_{1-}} & \frac{i}{\xi_{2+}} & \frac{1}{\xi_{2-}} \\ \frac{-\cos\theta_r^{\mathrm{I}}}{\xi_{1+}} & \frac{-i\cos\theta_r^{\mathrm{II}}}{\xi_{1-}} & \frac{\cos\theta_t^{\mathrm{I}}}{\xi_{2+}} & \frac{i\cos\theta_t^{\mathrm{II}}}{\xi_{2-}} \end{pmatrix}, \tag{5.45b}$$

$$\overline{D}_i = \begin{pmatrix} \frac{\eta_1}{\xi_{1+}} \\ \frac{i\cos\theta_i^{\mathrm{I}}\eta_1}{\xi_{1+}} \\ \frac{-i}{\xi_{1+}} \\ \frac{\cos\theta_i^{\mathrm{I}}}{\xi_{1+}} \end{pmatrix} D_i^{\mathrm{I}}. \tag{5.45c}$$

In the medium in case 1 with $\varepsilon > 0$, $\mu > 0$ and $\varepsilon\mu > \xi^2$, both the type-I and type-II waves exist for the reflected fields. In the medium in case 3 with $\varepsilon < 0$, $\mu < 0$ and $\varepsilon\mu > \xi^2$, both the type-I and type-II waves exist for the transmitted fields. Due to the chirality parameter, the two reflection angles are different, and the two refraction angles are also different. These are in sharp contrast with the waves in isotropic and anisotropic media. Besides, since k_2^{I} and k_2^{II} are both positive, negative refractions occur for both the type-I and type-II waves, and the refraction angles θ_t^{I} and θ_t^{II} satisfy

$$\left(\sqrt{\varepsilon_1\mu_1} + \xi_1\right)\sin\theta_i^{\mathrm{I}} = \left(\sqrt{\varepsilon_2\mu_2} + \xi_2\right)\sin\theta_t^{\mathrm{I}} = \left(\sqrt{\varepsilon_2\mu_2} - \xi_2\right)\sin\theta_t^{\mathrm{II}}, \tag{5.46}$$

as shown in Fig. 5.1. The existence of the solutions validates the negative refraction in biisotropic chiral metamaterials.

In applications people are more interested in the negative refraction that a wave is incident from air. For convenience we use the k surfaces to intuitively show that negative refraction can occur at the air–medium interface. We limit our discussion to the incidence of a type-I wave from air to a biisotropic chiral medium. The incident plane is the xz plane, and the boundary between the air and the medium is the yz plane. In this geometry the three-dimensional k space reduces to a two-dimensional plane. According to the dispersion relation in Eq. (3.24), the wave vector components satisfy

$$\frac{k_x^2}{\varepsilon_0\mu_0} + \frac{k_z^2}{\varepsilon_0\mu_0} = \omega^2 \tag{5.47a}$$

in the air, while in the biisotropic chiral medium, according to Eqs. (5.25, 5.27), the wave vector components satisfy

$$\frac{k_x^2}{\left(\sqrt{\varepsilon\mu} + \xi\right)^2} + \frac{k_z^2}{\left(\sqrt{\varepsilon\mu} + \xi\right)^2} = \omega^2 \tag{5.47b}$$

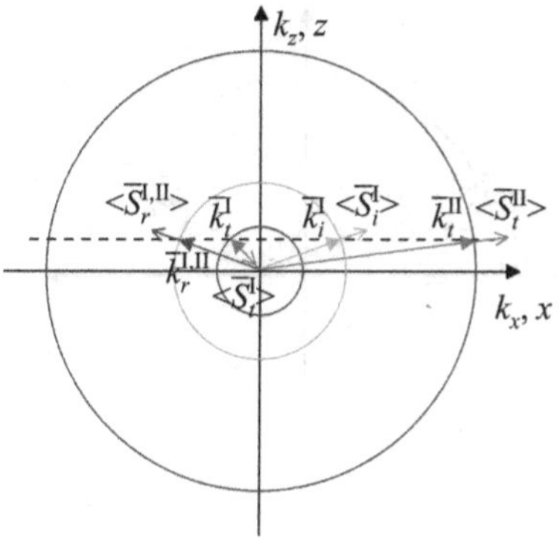

Figure 5.2 k surfaces for the waves in the air and in a biisotropic chiral medium, where the biisotropic chiral medium is a medium in case 2 with $\varepsilon > 0$, $\mu > 0$ and $\varepsilon\mu < \xi^2$.

for a type-I wave with $\varepsilon\mu > \xi^2$ and a type-II wave with $\varepsilon\mu < \xi^2$, and

$$\frac{k_x^2}{\left(\sqrt{\varepsilon\mu} - \xi\right)^2} + \frac{k_z^2}{\left(\sqrt{\varepsilon\mu} - \xi\right)^2} = \omega^2 \tag{5.47c}$$

for a type-I wave with $\varepsilon\mu < \xi^2$ and a type-II wave with $\varepsilon\mu > \xi^2$. We can see that the k surfaces in the air and in the biisotropic chiral medium are both circles. In the following we show some examples of negative refraction using the k surfaces.

1. The biisotropic chiral medium is a medium in case 2 with $\varepsilon > 0$, $\mu > 0$ and $\varepsilon\mu < \xi^2$. In this example negative refraction is expected to occur for only the type-I wave. For the incident field, its wave vector k_i^{I} is in the first quadrant, as shown in Fig. 5.2. According to the continuity condition of the wave vectors in Eq. (5.43), the wave vectors of the reflected fields $k_r^{\mathrm{I,II}}$ are in the second quadrant. For the transmitted fields, according to Tabs. 5.1 and 5.2, the directions of the wave vectors for the type-I and type-II waves are $\hat{e}_3$ and $-\hat{e}_3$, respectively. The directions of the Poynting's vectors are both $-\hat{e}_3$. Thus k_t^{I} must be in the second quadrant and k_t^{II} must be in the first quadrant, so that the Poynting's vectors are pointing away from the interface, as shown in Fig. 5.2. Negative refraction occurs because of the negative refraction angle of the energy flow.

2. The biisotropic chiral medium is a medium in case 3 with $\varepsilon < 0$, $\mu < 0$ and $\varepsilon\mu > \xi^2$. In this example negative refraction is expected to occur for both the type-I and type-II waves. For the incident field, its wave vector k_i^{I} is in the first quadrant, as shown in Fig. 5.3. According to the continuity condition of the wave vectors, the wave vectors of the reflected fields $k_r^{\mathrm{I,II}}$ are in the second

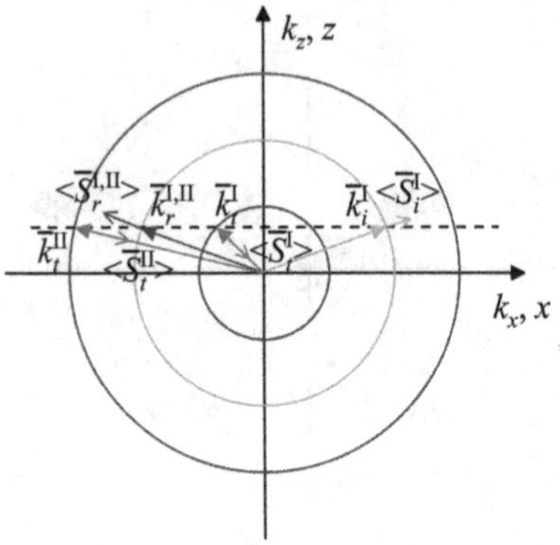

Figure 5.3 k surfaces for the waves in the air and in a biisotropic chiral medium, where the biisotropic chiral medium is a medium in case 3 with $\varepsilon < 0$, $\mu < 0$ and $\varepsilon\mu > \xi^2$.

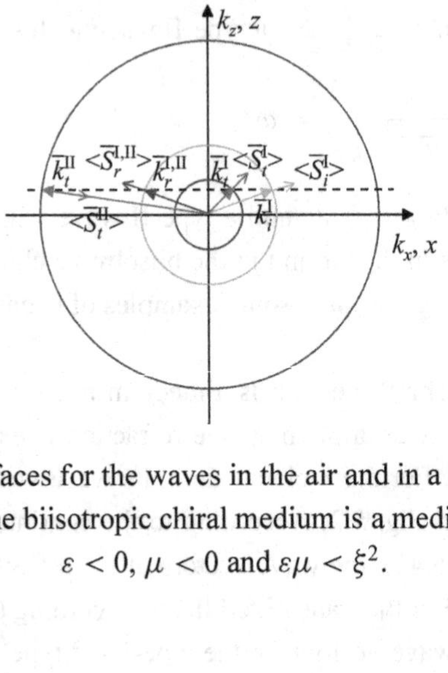

Figure 5.4 k surfaces for the waves in the air and in a biisotropic chiral medium, where the biisotropic chiral medium is a medium in case 4 with $\varepsilon < 0$, $\mu < 0$ and $\varepsilon\mu < \xi^2$.

quadrant. For the transmitted fields, according to Tabs. 5.1 and 5.2, the directions of the wave vectors are $\hat{e}_3$. The directions of the Poynting's vectors are $-\hat{e}_3$. Thus $k_t^{\mathrm{I,II}}$ must be in the second quadrant so that the Poynting's vector is pointing away from the interface, as shown in Fig. 5.3. Negative refraction occurs because of the negative refraction angle of the energy flow.

3. The biisotropic chiral medium is a medium in case 4 with $\varepsilon < 0$, $\mu < 0$ and $\varepsilon\mu < \xi^2$. In this example negative refraction is expected to occur for only the type-II waves. For the incident field, its wave vector k_i^{I} is in the first quadrant, as shown in Fig. 5.4. According to the continuity condition of the wave vectors, the wave vectors of the reflected fields $k_r^{\mathrm{I,II}}$ are in the second quadrant.

For the transmitted fields, according to Tabs. 5.1 and 5.2, the directions of the wave vectors for the type-I and type-II waves are $\hat{e}_3$ and $-\hat{e}_3$, respectively. The directions of the Poynting's vectors are $\hat{e}_3$. Thus k_t^{I} must be in the first quadrant and k_t^{II} must be in the second quadrant, so that the Poynting's vector is pointing away from the interface, as shown in Fig. 5.4. Negative refraction occurs because of the negative refraction angle of the energy flow.

6 Experimental Implementations and Applications

6.1 Introduction

In this section we introduce the experimental implementations of various kinds of metamaterials. Although isotropic metamaterials have the simplest constitutive parameters, their experimental implementations are quite challenging compared with those of other kinds of metamaterials. Considering the degree of fabrication difficulties, we start from anisotropic metamaterials, then introduce bianisotropic metamaterials, chiral metamaterials and isotropic metamaterials.

6.2 Experimental Implementations of Anisotropic Metamaterials

To achieve anisotropic metamaterials, we only need to design the electric and/or magnetic responses for designated polarizations. Thus the difficulty in practice is largely reduced compared with the isotropic ones. The experimental verifications are widely demonstrated in the microwave region [10–12]. The negative-index metamaterial consists of a two-dimensionally periodic array of copper SRRs and wires, fabricated by shadow mask/etching technology [10]. In this design, SRRs are used to produce negative magnetic permeability over a particular frequency region, and wire elements offer negative electric permittivity in an overlapping frequency region. The phenomenon of negative refraction has been observed in an experiment by measuring the scattering angle of the transmitted beam through a prism fabricated from this material. In the terahertz region, magnetic response has been achieved in a planar structure composed of nonmagnetic conductive resonant elements [13]. The magnetic response is due to the constituent SRRs, and it scales with dimensions in accordance with Maxwell's equations. The scalability of these magnetic metamaterials promises their applications into the higher terahertz range [14]. Pushing the operating frequency deeper to reach ultimately optical frequencies is important because no natural magnetic materials exist at such high frequencies. At low frequencies, the magnetic resonant frequency scales reciprocally with the structural size. At high frequencies, however, the linear scaling breaks down [15]. The breakdown of linear scaling originates from the kinetic energy of the electrons in the metal, which cannot be neglected anymore in comparison with the magnetic energy at high frequencies. Above the linear scaling regime, the resonant frequency of SRR saturates and ultimately ceases to reach a negative value. The highest possible resonant frequency exhibiting negative permeability strongly increases with the number of cuts in SRR.

To achieve negative refraction in optical frequencies, researchers have proposed fishnet structures with both negative permittivity and permeability [16, 17]. A 3-D array of metal wires aligned with the polarization direction of the incident electric field serves as an effective medium with lowered volumetric plasma frequency. Wave propagation below this frequency is not allowed because of the negative permittivity. Another array of metal strips along the direction of the magnetic field induces antisymmetric conductive currents across the dielectric layers, giving rise to a magnetic bandgap. A propagation band with negative refractive index appears in the overlapped region of the forbidden gaps of the electric and magnetic bandgaps. The experimental realization of the cascaded fishnet structures is then demonstrated in the near-infrared regime [17]. Since most metamaterials with negative refractive index are based on metallic artificial atoms, the loss is inherently a major issue that limits their applications in practice, especially in the near-infrared and visible wavelength ranges. One way to address this issue is the incorporation of gain material in the high-local-field areas of a metamaterial [17]. Active optical negative-index metamaterials have been demonstrated by replacing the dielectric spacer with a gain layer of dye in fishnet structure.

6.3 Experimental Implementations of Bianisotropic Metamaterials

Experimental implementations of bianisotropic metamaterials have been demonstrated in microwave [18–20], far-infrared [21] and optical regimes [22–25]. Among these, split-ring resonator is the most representative structure for bianisotropy in which a large magnetic response can be achieved. Since the dimension of unit cell is in subwavelength scale, the optical response of the SRR can be analyzed via the quasi-static approach. We consider a SRR with one gap shown in Fig. 6.1, where the radius of the SRR is r_0 and the periodicities in the x, y and z directions are a_x, a_y and a_z, respectively. Assuming an external magnetic field polarized along the normal direction of the SRR plane with $\overline{B} = \hat{z}B_z^{\text{ext}} \exp(-i\omega t)$, a strong magnetic response is induced and there is a current in the metal due to the electromotive force (EMF). According to Lenz's law, the EMF at the gap is

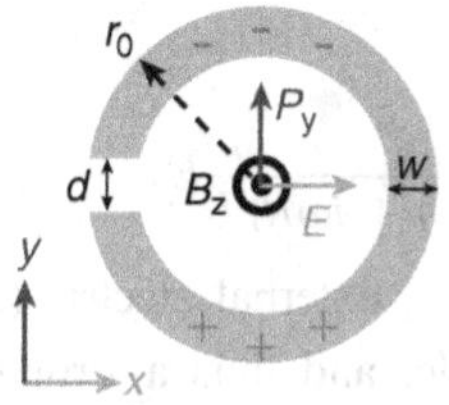

Figure 6.1 The bianisotropic effect in a SRR structure.

$$\text{EMF} = -\frac{\partial}{\partial t}\iint d\overline{S}\cdot\overline{B} = i\omega\pi r_0^2 B_z^{\text{ext}}. \tag{6.1}$$

The resonance of the SRR can be explained using an equivalent circuit model, which includes a resistor R from the metal loss, an inductor L of an almost closed loop and a capacitor C formed at the gap. Using Kirchhoff's voltage law, the motion of the electric current I can be described by

$$\text{EMF} = \oint d\bar{l}\cdot\overline{E} = RI + \frac{1}{-i\omega C}I - i\omega LI = i\omega\pi r_0^2 B_z^{\text{ext}}, \tag{6.2}$$

and subsequently we obtain the expression for the current

$$I = \frac{i\omega\pi r_0^2 B_z^{\text{ext}}}{R + 1/(-i\omega C) - i\omega L} = \frac{-\omega^2\pi r_0^2 L^{-1}}{\omega^2 - 1/(LC) + i\omega R/L}B_z^{\text{ext}}. \tag{6.3}$$

This leads to the magnetic dipole in the z direction

$$m_z = \pi r_0^2 I = \frac{-\omega^2\pi^2 r_0^4 L^{-1}}{\omega^2 - 1/(LC) + i\omega R/L}B_z^{\text{ext}}. \tag{6.4}$$

Meanwhile, the electric dipole is induced by the current at the gap with the moment given by the charges separated on the capacitor plates with $p(t) = d\int I(t)\,dt$. This leads to the electric dipole moment in the frequency domain

$$p_y = \frac{dI}{-i\omega} = \frac{-i\omega\pi r_0^2 L^{-1} d}{\omega^2 - 1/(LC) + i\omega R/L}B_z^{\text{ext}}. \tag{6.5}$$

Similarly, if we apply an external electric field in the y direction E_y, an electromotive voltage will be induced as well with $\Psi = E_y d$. The induced current in the SRR loop is thus obtained as

$$I = \frac{E_y d}{R + 1/(-i\omega C) - i\omega L}, \tag{6.6}$$

with the induced magnetic dipole

$$m_z = \pi r_0^2 I = \frac{\pi r_0^2 d}{R + 1/(-i\omega C) - i\omega L}E_y = \frac{i\omega\pi r_0^2 L^{-1} d}{\omega^2 - 1/(LC) + i\omega R/L}E_y, \tag{6.7}$$

and the electric dipole

$$p_y = \frac{dI}{-i\omega} = \frac{-L^{-1}d^2}{\omega^2 - 1/(LC) + i\omega R/L} E_y. \tag{6.8}$$

Notably, in the SRR structure, external electric (magnetic) field can induce the magnetic (electric) dipole, and thus a strong magnetoelectric coupling (or bianisotropic effect) exists. By homogenization of a medium we can get the macroscopic effective parameters of the array of SRR resonators. In the case of free space surrounded, the macroscopic electric polarization P_y is the product of the individual electric dipole moment times its number density $N/V_{\text{unit}} = 1/(a_x a_y a_z)$, which results in $P_y = p_y/V_{\text{unit}}$. The macroscopic magnetic moment density M_z can be calculated by $M_z = m_z/V_{\text{unit}}$ as well. When this is combined with the constitutive relations of a medium with $\overline{D} = \overline{\overline{\varepsilon}}E + \overline{\overline{\xi}}H$ and $\overline{B} = \overline{\overline{\zeta}}E + \overline{\overline{\mu}}H$, we thus get the effective constitutive parameters

$$\bar{\varepsilon} = \varepsilon_0 \begin{bmatrix} \varepsilon_x & 0 & 0 \\ 0 & \varepsilon_y & 0 \\ 0 & 0 & \varepsilon_z \end{bmatrix}, \tag{6.9a}$$

$$\bar{\mu} = \mu_0 \begin{bmatrix} \mu_x & 0 & 0 \\ 0 & \mu_y & 0 \\ 0 & 0 & \mu_z \end{bmatrix}, \tag{6.9b}$$

$$\bar{\xi} = \frac{1}{c} \begin{bmatrix} 0 & 0 & 0 \\ 0 & 0 & -i\xi_0 \\ 0 & 0 & 0 \end{bmatrix}, \tag{6.9c}$$

$$\bar{\zeta} = \frac{1}{c} \begin{bmatrix} 0 & 0 & 0 \\ 0 & 0 & 0 \\ 0 & i\xi_0 & 0 \end{bmatrix}, \tag{6.9d}$$

where c is the free-space velocity of light. The bianisotropic effect of the SRR can also be analyzed by Babinet principles from which we can obtain the relationship of the electromagnetic responses between complementary metamaterials [26, 27].

The retrieval method for such bianisotropic metamaterials is significantly different from that for ordinary metamaterials. Various retrieval strategies have been discussed theoretically in the past few years [19, 25, 28–31]. Here, we briefly introduce the retrieval procedure for the aforementioned bianisotropic metamaterial. From Eqs. (6.9a, 6.9b, 6.9c, 6.9d) we can see that there are seven unknown quantities, ε_x, ε_y, ε_z, μ_x, μ_y, μ_z and ξ_0, to be determined. Specifically, ε_x, ε_z, μ_x and μ_y can be obtained from the ordinary retrieval method for isotropic medium, while the bianisotropic term does not contribute to the optical response. A different situation occurs when a plane wave propagates along

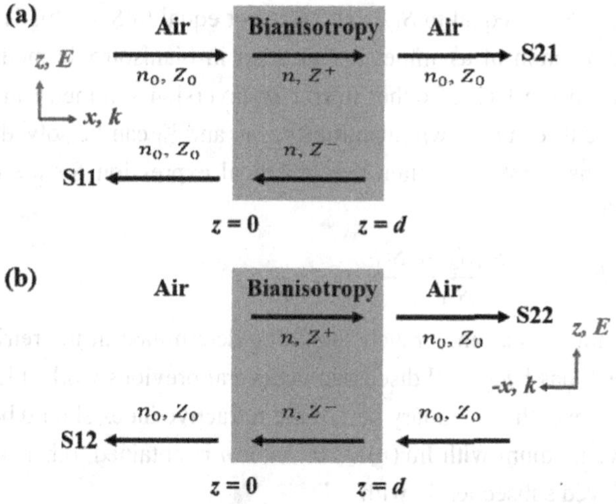

Figure 6.2 Schematics of a homogeneous bianisotropic slab placed in free space for the calculation of S parameters. (a) A plane wave propagates in the $+x$ direction. (b) A plane wave propagates in the $-x$ direction.

the $+x$ direction with the electric field polarized in the y direction. In this case the parameters of ε_y, μ_z and ξ_0 will be active and the other four parameters are not involved. This is the case of bianisotropic response in which the traditional retrieval method for isotropic medium is not applicable. At this point, the characteristic impedances have different values for plane waves propagating in the two opposite directions of the x axis with $z^{\pm} = \mu_z/(n \pm i\xi_0)$, where $n = \sqrt{\varepsilon_y \mu_z - \xi_0^2}$ is the effective refractive index, and the signs of '+' and '-' represent the waves traveling in the + and $-x$ directions, respectively. For a passive medium, the real part of the impedance should be positive with $\mathrm{Re}\left(z^{\pm}\right) \geq 0$. Fig. 6.2 shows the schematics of a bianisotropic slab placed in free space. By applying the boundary conditions, it is easy to obtain the expressions of the reflection and transmission coefficients with

$$S_{11} = \frac{i2\sin(nk_0 d)\left[(\xi_0 + i\mu_z)^2 + n^2\right]}{\left[\xi_0^2 + (\mu_z + n)^2\right]e^{-in)k_0 d} - \left[\xi_0^2 + (\mu_z - n)^2\right]e^{ink_0 d}}, \tag{6.10a}$$

$$S_{21} = S_{12} = \frac{4\mu_z n}{\left[\xi_0^2 + (\mu_z + n)^2\right]e^{-ink_0 d} - \left[\xi_0^2 + (\mu_z - n)^2\right]e^{ink_0 d}}, \tag{6.10b}$$

$$S_{22} = \frac{i2\sin(nk_0 d)\left[(\xi_0 - i\mu_z)^2 + n^2\right]}{\left[\xi_0^2 + (\mu_z + n)^2\right]e^{-ink_0 d} - \left[\xi_0^2 + (\mu_z - n)^2\right]e^{ink_0 d}}. \tag{6.10c}$$

It is clear that S_{21} is equal to S_{12}, yet S_{11} is not equal to S_{22}. This is the basic optical phenomenon of asymmetric reflection in bianisotropic media, which originates from the lack of either mirror or inversion symmetry in the SRR structure. The three unknown quantities ε_y, μ_z and ξ_0 can be solved from the three equations. First, we obtain the analytical expression for the refractive index n with

$$\cos(nk_0 d) = \frac{1 - S_{11}S_{22} + S_{21}^2}{2S_{21}}. \tag{6.11}$$

It is worth noting that one branch should be determined in the retrieval procedure, which has been well discussed in several previous works [32, 33]. As mentioned above, the imaginary part of the refractive index should be positive for a passive medium with $\mathrm{Im}(n) \geq 0$. After n is obtained, other parameters can be retrieved subsequently with

$$\xi_0 = \frac{n}{-2\sin(nk_0 d)} \frac{S_{11} - S_{22}}{S_{21}}, \tag{6.12a}$$

$$\mu_z = \frac{in}{\sin(nk_0 d)} \left[\frac{2 + S_{11} + S_{22}}{2S_{21}} - \cos(nk_0 d) \right], \tag{6.12b}$$

$$\varepsilon_y = \frac{n^2 + \xi_0^2}{\mu_z}. \tag{6.12c}$$

So far, all of the constitutive parameters of the bianisotropic SRR structure are retrieved.

6.4 Experimental Implementations of Chiral Metamaterials

Chiral metamaterials are a subset of metamaterials in which the mirror symmetry should always be broken. Chiral metamaterials also exhibit a number of intriguing properties. For instance, if the chirality is sufficiently strong, negative refractive indices can be achieved even though neither ε nor μ is negative. This concept was theoretically proposed by Tretyakov et al. [34] and discussed by Pendry [35] and Monzon et al. [36]. Experimental realizations of chiral metamaterials have been demonstrated in the microwave [37–39], terahertz [40] and optical regimes [41, 42]. In addition, chiral metamaterials could be designed to achieve strong optical activity [43], large circular dichroism [44], asymmetric transmission [45, 46] and superchiral field [47, 48].

As we know, a material is defined as chiral if it lacks any plane of mirror symmetry. Hence the key point in the realization of chiral metamaterials is the breaking of the mirror symmetry. Two major strategies are widely adopted toward this purpose, where one is constructing three-dimensional chiral structures and the other is stacking nonchiral metamaterial layers together with certain twisted angles. In the optical regime one classic example of the former

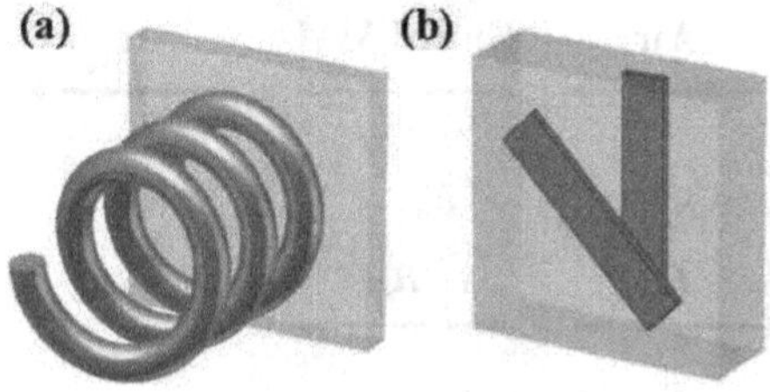

Figure 6.3 Artificial chiral metaatoms. (a) A three-dimensional (3-D) helix. (b) A stacked planar chiral structure.

strategy is a three-dimensional helix structure as shown in Fig. 6.3a. A uniaxial photonic metamaterial composed of three-dimensional gold helices has been demonstrated in Ref. [44], where a broadband circular dichroism has been achieved. In practice, top-down fabrication methods for three-dimensional helices include direct laser writing, electrochemical deposition, electroless silver plating, glancing-angle deposition, colloidal hole-mask lithography, on-edge lithography and so on [49]. In addition, bottom-up fabrication methods such as self-assembly techniques have also been adopted in the construction of three-dimensional chiral bulks. Top-down approaches often provide excellent fabrication accuracy and flexibility, yet they are normally costly, low-speed and cannot be applied to massive production. The situation is almost reversed in the case of the bottom-up approaches, which are cheaper, faster and regarded as the best methods for mimicking the self-assembly process in real life [49]. For the second strategy of stacked-planar chiral metamaterials as shown in Fig. 6.3b, the structures are generally fabricated by standard electro-beam lithography along with alignment in a layer-by-layer manner. Various types of stacked chiral metamaterials have been experimentally demonstrated, such as twisted split rings [50], chiral oligomers [51], L-shape [52] and twisted-arc chiral nanostructures [53].

Parameter retrieval is a necessary and essential procedure of obtaining the macroscopic electromagnetic parameters of a medium. It is usually based on the transmission and reflection coefficients, namely the S parameters, obtained from either simulations or measurements, as shown in Fig. 6.4. For a biisotropic chiral medium, the electromagnetic response is usually described by the permittivity ε, the permeability μ and the chirality parameter χ. All of them are intrinsic properties of the material and irrelevant to the thickness of the slab. However, for metamaterial slabs, the smallest slab thickness is limited by the size of the unit cell, and therefore the retrieval solution would be generally multibranched. In the retrieval process the branches need to be carefully chosen

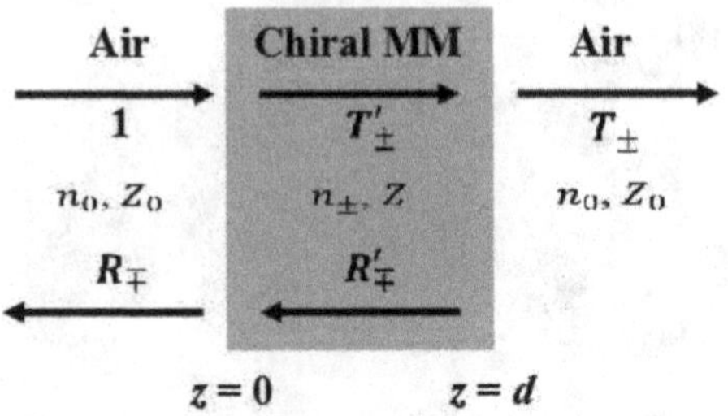

Figure 6.4 Schematics of the transmission and reflection of a free-standing chiral metamaterial slab.

in order to obey the energy conservation rules. For a chiral slab the refractive indices for right- (+) and left-handed (−) circular polarizations are differently given by $n_{\pm} = \sqrt{\varepsilon\mu} \pm \xi$, with the corresponding wave vectors $k_{\pm} = k_0 n_{\pm}$. Waves with opposite handedness will obtain different accumulations of phase as they travel through a chiral material. However, the two circular polarizations have the same wave impedance with $Z = Z_0\sqrt{\mu/\varepsilon}$, where $Z_0 = \sqrt{\mu_0/\varepsilon_0}$ is the free-space impedance. The normalized impedance can then be defined as $z_{\text{norm}} = Z/Z_0$. Now we apply the boundary conditions of continuous electric and magnetic fields at $z = 0$ and $z = d$ with

$$1 + R_{\mp} = T'_{\pm} + R'_{\mp}, \tag{6.13a}$$

$$1 - R_{\mp} = \frac{T'_{\pm} - R'_{\mp}}{z}, \tag{6.13b}$$

$$T'_{\pm}e^{ik_{\pm}d} + R'_{\mp}e^{-ik_{\mp}d} = T_{\pm}, \tag{6.13c}$$

$$\frac{T'_{\pm}e^{ik_{\pm}d} - R'_{\mp}e^{-ik_{\mp}d}}{z} = T_{\pm}. \tag{6.13d}$$

Then we can get the transmission and reflection coefficients with

$$T_{\pm} = \frac{4ze^{ik_{\pm}d}}{(1+z)^2 - (1-z)^2\, e^{i(n_+ + n_-)k_0 d}}, \tag{6.14a}$$

$$R_+ = R_- = \frac{(1-z)^2\left(e^{i(n_+ + n_-)k_0 d} - 1\right)}{(1+z)^2 - (1-z)^2\, e^{i(n_+ + n_-)k_0 d}}. \tag{6.14b}$$

It is worth noting that the reflection coefficients are identical for two polarizations and therefore we can define $R_{\pm} = R$. Then, from the reflection and transmission coefficients, we can get

$$z = \pm\sqrt{\frac{(1+R)^2 - T_+T_-}{(1-R)^2 - T_+T_-}}, \tag{6.15a}$$

$$n_{\pm} = \frac{1}{k_0 d}\left\{ i \ln\left[\frac{1}{T_{\pm}}\left(1 + \frac{1-z}{1+z}R\right)\right] \pm 2m\pi \right\}, \tag{6.15b}$$

where m is an integer determined by the branches. The retrieved results must obey the conditions for a passive medium with $\mathrm{Re}(z) > 0$ and $\mathrm{Im}\,(n_{\pm}) > 0$.

The approach of parameter retrieval requires the tensor format of the constitutive parameters to be determined in advance. This feature would dramatically increase the complexity if sophisticated optical effects are involved, such as anisotropy, nonlocality and spatial dispersion [54, 55]. Symmetry consideration can offer us a convenient route to directly analyze the optical response of various metamaterial structures. For planar media the optical responses can be described by a Jones matrix [56, 57], which relates the complex amplitudes of the incident and scattered fields with

$$\begin{pmatrix} E_r^x \\ E_r^y \end{pmatrix} = \begin{pmatrix} r_{xx} & r_{xy} \\ r_{yx} & r_{yy} \end{pmatrix}\begin{pmatrix} E_i^x \\ E_i^y \end{pmatrix} = \overline{\overline{R}}\begin{pmatrix} E_i^x \\ E_i^y \end{pmatrix}, \tag{6.16a}$$

$$\begin{pmatrix} E_t^x \\ E_t^y \end{pmatrix} = \begin{pmatrix} t_{xx} & t_{xy} \\ t_{yx} & t_{yy} \end{pmatrix}\begin{pmatrix} E_i^x \\ E_i^y \end{pmatrix} = \overline{\overline{T}}\begin{pmatrix} E_i^x \\ E_i^y \end{pmatrix}. \tag{6.16b}$$

Here $\overline{\overline{R}}$ and $\overline{\overline{T}}$ are called the reflection and transmission matrices for a linear polarization, respectively. And E_i^x, E_r^x and E_t^x are the incident, reflected and transmitted electric fields in the x direction, respectively. The similar notations with a subscript y represent the electric fields in the y direction. By transformation of the bases, the Jones matrices for a circular polarization can be obtained as

$$\begin{aligned} \overline{\overline{R}}_{\text{circ}} &= \begin{pmatrix} r_{++} & r_{+-} \\ r_{-+} & r_{--} \end{pmatrix} = \overline{\overline{\Lambda}}^{-1}\overline{\overline{R\Lambda}} \\ &= \frac{1}{2}\begin{pmatrix} r_{xx} + r_{yy} + i\left(r_{xy} - r_{yx}\right) & r_{xx} - r_{yy} - i\left(r_{xy} + r_{yx}\right) \\ r_{xx} - r_{yy} + i\left(r_{xy} + r_{yx}\right) & r_{xx} + r_{yy} - i\left(r_{xy} - r_{yx}\right) \end{pmatrix}, \end{aligned} \tag{6.17a}$$

$$\begin{aligned} \overline{\overline{T}}_{\text{circ}} &= \begin{pmatrix} t_{++} & t_{+-} \\ t_{-+} & t_{--} \end{pmatrix} = \overline{\overline{\Lambda}}^{-1}\overline{\overline{T\Lambda}} \\ &= \frac{1}{2}\begin{pmatrix} t_{xx} + t_{yy} + i\left(t_{xy} - t_{yx}\right) & t_{xx} - t_{yy} - i\left(t_{xy} + t_{yx}\right) \\ t_{xx} - t_{yy} + i\left(t_{xy} + t_{yx}\right) & t_{xx} + t_{yy} - i\left(t_{xy} - t_{yx}\right) \end{pmatrix}, \end{aligned} \tag{6.17b}$$

where

$$\overline{\overline{\Lambda}} = \frac{1}{\sqrt{2}}\begin{pmatrix} 1 & 1 \\ i & -i \end{pmatrix} \tag{6.18}$$

is the change of basis matrix, and the subscript + (−) denotes clockwise (counterclockwise) circularly polarized waves as viewed along the $+z$ direction, respectively.

Next we introduce how to apply symmetry considerations to predict the optical properties of a metamaterial structure [49]. If a metamaterial exhibits a certain symmetry group, the Jones matrices after transformation must be identical to the original ones. For the case of a mirror symmetry with respect to the incident plane, which is defined as the xz plane, the off-diagonal elements in the Jones matrices of a linear polarization must eliminate with

$$r_{xy} = r_{yx} = t_{xy} = t_{yx} = 0, \tag{6.19}$$

and thus the Jones matrices for a circular polarization become symmetric with

$$r_{++} = r_{--}, \tag{6.20a}$$

$$r_{+-} = r_{-+}, \tag{6.20b}$$

$$t_{++} = t_{--}, \tag{6.20c}$$

$$t_{+-} = t_{-+} \tag{6.20d}$$

[57]. Since the optical activity is usually characterized by

$$\theta = \frac{\arg(t_{++}) - \arg(t_{--})}{2}, \tag{6.21}$$

there is no polarization rotation due to chirality. The off-diagonal elements are also identical so that two circularly polarized waves have the same efficiency of polarization conversion. As a result, neither optical activity nor circular dichroism can exist in mirror-symmetric structures. This result explains, from the point view of the Jones matrix, why chirality exists only in structures that lack mirror symmetries.

In the cases of threefold (C_3) or fourfold (C_4) rotational symmetries, the symmetry consideration results in $r_{xx} = r_{yy}$ and $r_{xy} = -r_{yx}$ [57]. For circular polarizations we therefore get $r_{+-} = r_{-+} = 0$. The same reasoning holds true for the transmission matrix so that $t_{+-} = t_{-+} = 0$. In the case of normal incidence, the De Hoop reciprocity indicates that the reflection matrix must obey the general identity $\overline{\overline{R}} = \overline{\overline{R}}^T$, where the superscript T represents the transpose operation [58–60]. If we combine these restrictions, the linear polarization conversions are derived to be zero with $r_{xy} = r_{yx} = 0$. Consequently, the Jones matrices can be expressed as

$$\overline{\overline{R}}_{\text{circ}} = \begin{pmatrix} r_{xx} & 0 \\ 0 & r_{xx} \end{pmatrix} = r_{xx}\overline{\overline{I}}, \tag{6.22}$$

$$\overline{\overline{T}}_{\text{circ}} = \begin{pmatrix} t_{xx} + it_{xy} & 0 \\ 0 & t_{xx} - it_{xy} \end{pmatrix}, \tag{6.23}$$

where $\bar{\bar{I}}$ is the identity matrix. Therefore, the reflection coefficient is the same for all polarizations, which is also predicted by the effective medium method. For a structure with either C_3 or C_4 symmetry, the chirality parameter can be simplified to a scalar χ, which leads to the absence of polarization conversions [61–63]. Moreover, calculation shows that the reflections are identical for both the left-handed circular polarization and right-handed circular polarization illuminations in an isotropic chiral material [64]. The underlying reason is that the impedance of this chiral medium is only determined by the ratio between its permittivity and permeability and is independent of the chirality parameter. Furthermore, when losses are absent, analyses in reciprocity and energy conservation indicate that the transmission coefficients of two spin states are identical as well [60]. If the chiral metamaterials become anisotropic, such as the twofold (C_2) rotational symmetry, the polarization conversion between the two spin states in reflection would appear. This feature has been applied in the design of chiral metamirror, which behaves as a highly efficient circular polarizer in reflection without handedness reversal [65].

6.5 Experimental Implementations of Isotropic Metamaterials

To achieve an isotropic metamaterial, it is straightforward to start from a cubic unit cell. The SRR and the continuous wire offer a resonant negative permeability and a negative permittivity, respectively. However, this occurs only when the magnetic field is normal to the SRR plane and the electric field is parallel to the wire. Therefore a pair of SRR and wire can provide the negative refraction in a particular spatial direction. A simple approach for isotropic responses is the arrangement of the artificial resonant elements into a highly symmetric structure. There are some design criteria to attain the best possible isotropy [66]:

(1) To avoid the coupling of electric field to the magnetic resonance, we have to provide mirror symmetry of the SRR plane with respect to the direction of the electric field.

(2) For highly efficient suppression of the cross-polarization scattering amplitudes, the inversion symmetry of the unit cell should not be broken in both directions that are perpendicular to the direction of propagation. Therefore it would be better to center the SRRs and the wires on the faces of the unit cell.

(3) The description in terms of a homogeneous effective medium is possible only if the isotropy in the direction of the finite dimension is preserved. Therefore we have to set the second surface of the slab as a repetition of the opposite face.

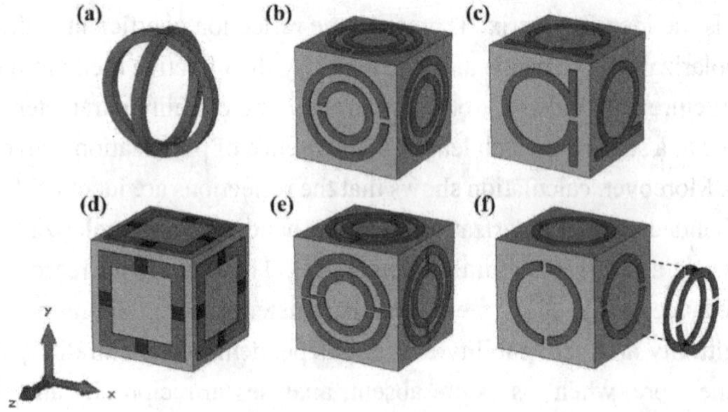

Figure 6.5 Cubic constitutive elements for isotropic metamaterials. (a) Crossed SRRs. (b) Planar SRRs. (c) Planar omega particles. (d) Four-gap SRRs. (e) Nonbianisotropic SRRs. (f) Broadside-coupled double-slit SRRs.

(4) Keeping the magnetic resonant frequency well below the first periodicity bandgap to avoid the effects of periodicity.

(5) Aligning the wires to the middle of the SRRs can minimize the disturbance between them. It is also favorable not to put the edges of the SRRs too close to each other.

A first attempt to design an isotropic magnetic metamaterial was first proposed in 2002 [67]. In that work a spherical magnetic resonator is formed by two SRRs crossed together, where an isotropic electromagnetic response can be achieved in two dimensions (Fig. 6.5a). To reduce the fabrication difficulty in practice, researchers have also done work on the design of cubic units [68, 69]. SRR (shown in Fig. 6.5b) or omega (shown in Fig. 6.5c) particles are arranged in each face of a cubic unit to ensure invariance under a cubic symmetry. If only the magnetic and electric dipole representations of the SRRs and omega particles are considered, these arrangements are invariant under cubic symmetries. However, this invariant is not enough to guarantee the isotropic behaviors due to the coupling between the planar resonators [70].

The isotropic metamaterial designed with full invariance under the whole group of symmetry of the cube was first proposed and simulated in [66]. A four-gap SRR is selected as the magnetic resonant element, as shown in Fig. 6.5(d). The gap width has to be significantly reduced in comparison with a single-gap SRR due to the receded capacitance. Alternately, one could increase the dielectric constant inside the gap to move the resonant frequency down. However, this design is unfortunately very difficult to implement in practice because it cannot be manufactured by the standard photoetching techniques, and the gaps of the SRR have to be filled with a high-relative-permittivity dielectric. To reduce the difficulty in fabrication, researchers have also demonstrated isotropic electromagnetic responses in metamaterials by using spatial symmetries [70]. By arranging nonbianisotropic SRRs in a cubic unit as shown in Fig. 6.5e, we can achieve invariance under most symmetries without the inversion. In this case the whole structure may show biisotropy. A chiral behavior has been clearly observed in the experiment, in which the cross-polarization transmission coefficient is measured in a rectangular waveguide. If chirality is not desired, the design must be invariant under the whole group of symmetry, and the inversion symmetry needs to be included. This can be achieved by a cubic array of broadside-coupled double-slit SRRs as shown in Fig. 6.5f.

Another group of attempts in isotropic metamaterials is based on the high-index inclusions [71, 72]. In these works lattices of dielectric or paramagnetic spheres with high refractive indices are proposed. When the refractive index is high enough, the internal wavelength becomes relatively small with respect to the wavelength in free space, and Mie resonances of the spheres turn up and produce the negative permittivity or permeability. The isotropy is ensured by the spherical shapes of the metaatoms. This strategy often suffers from two drawbacks, where one is the difficulty in the fabrication of high-index inclusions with low loss and the other is that Mie-resonance-based responses often provide narrow bands. In conclusion, the systematic approach to design isotropic metamaterials is constructing metacrystals with required symmetry groups and obtaining the effective medium parameters from homogenization according to casualty laws.

7 Concluding Remarks

In this Element, we have introduced the basic theoretical concepts and design principles of negative refraction in metamaterials, as well as the experimental implementations of metamaterials. The uniaxial and biisotropic chiral metamaterials are taken as typical examples of anisotropic and bianisotropic metamaterials, respectively. The theoretical methods can be used to study other kinds of metamaterials, and we expect that more interesting negative refraction–related phenomena would be found in the near future. In particular, the *kDB* coordinate system that we used in this Element provides a simple way to describe wave propagation behavior in complicated media, and it will be very helpful to the study of negative refraction. The concept of metamaterials and their experimental implementations have not only extended human cognition of the universe but also shed light on some inconceivable applications.

References

1. Maxwell JC. A treatise on electricity and magnetism. London: Constable and Company; 1873.
2. Marder MP. Condensed matter physics. 2nd ed. New Jersey: John Wiley & Sons, Inc; 2010.
3. Rothwell EJ, Cloud MJ. Electromagnetics. New York: CRC Press; 2008.
4. Kong JA. Electromagnetic wave theory. Cambridge: Wiley and Sons, EMW Publishing; 2008.
5. Cui TJ, Smith DR, Liu R. Metamaterials – theory, design, and applications. New York: Springer; 2010.
6. Pendry JB, Holden AJ, Stewart WJ, Youngs I. Extremely low frequency plasmons in metallic mesostructures. Phys Rev Lett. 1996 Jun;76:4773–4776.
7. Maslovski SI, Tretyakov SA, Belov PA. Wire media with negative effective permittivity: A quasi-static model. Microwave and Optical Technology Letters. 2002;35(1):47–51.
8. Pendry JB, Holden AJ, Robbins DJ, Stewart WJ. Magnetism from conductors and enhanced nonlinear phenomena. IEEE Transactions on Microwave Theory and Techniques. 1999 Nov;47(11):2075–2084.
9. O'Brien S, Pendry JB. Magnetic activity at infrared frequencies in structured metallic photonic crystals. Journal of Physics: Condensed Matter. 2002;14(25):6383.
10. Shelby RA, Smith DR, Schultz S. Experimental verification of a negative index of refraction. Science. 2001;292(5514):77–79.
11. Smith DR, Padilla WJ, Vier DC, Nemat-Nasser SC, Schultz S. Composite medium with simultaneously negative permeability and permittivity. Phys Rev Lett. 2000 May;84:4184–4187.
12. Chen H, Wang Z, Zhang R, Wang H, Lin S, Yu F, et al. A metasubstrate to enhance the bandwidth of metamaterials. Scientific Reports. 2014;4:5264.
13. Yen TJ, Padilla WJ, Fang N, Vier DC, Smith DR, Pendry JB, et al. Terahertz magnetic response from artificial materials. Science. 2004;303(5663):1494–1496.
14. Linden S, Enkrich C, Wegener M, Zhou J, Koschny T, Soukoulis CM. Magnetic response of metamaterials at 100 terahertz. Science. 2004;306(5700):1351–1353.

15. Zhou J, Koschny T, Kafesaki M, Economou EN, Pendry JB, Soukoulis CM. Saturation of the magnetic response of split-ring resonators at optical frequencies. Phys Rev Lett. 2005 Nov;95:223902.
16. Zhang S, Fan W, Malloy KJ, Brueck SRJ, Panoiu NC, Osgood RM. Near-infrared double negative metamaterials. Opt Express. 2005 Jun;13(13):4922–4930.
17. Valentine J, Zhang S, Zentgra T, Ulin-Avila E, Genov DA, Bartal G, et al. Three-dimensional optical metamaterial with a negative refractive index. Nature. 2008;455:376.
18. Marqués R, Medina F, Rafii-El-Idrissi R. Role of bianisotropy in negative permeability and left-handed metamaterials. Phys Rev B. 2002 Apr;65:144440.
19. Chen X, Wu BI, Kong JA, Grzegorczyk TM. Retrieval of the effective constitutive parameters of bianisotropic metamaterials. Phys Rev E. 2005 Apr;71:046610.
20. Smith DR, Gollub J, Mock JJ, Padilla WJ, Schurig D. Calculation and measurement of bianisotropy in a split ring resonator metamaterial. Journal of Applied Physics. 2006;100(2):024507.
21. Xu X, Quan B, Gu C, Wang L. Bianisotropic response of microfabricated metamaterials in the terahertz region. J Opt Soc Am B. 2006 Jun;23(6):1174–1180.
22. Rill MS, Plet C, Thiel M, Staude I, von Freymann G, Linden S, et al. Photonic metamaterials by direct laser writing and silver chemical vapour deposition. Nature Materials. 2008;7:543–546.
23. Rill MS, Kriegler CE, Thiel M, von Freymann G, Linden S, Wegener M. Negative-index bianisotropic photonic metamaterial fabricated by direct laser writing and silver shadow evaporation. Opt Lett. 2009 Jan;34(1):19–21.
24. Kraft M, Braun A, Luo Y, Maier SA, Pendry JB. Bianisotropy and magnetism in plasmonic gratings. ACS Photonics. 2016;3(5):764–769.
25. Kriegler CE, Rill MS, Linden S, Wegener M. Bianisotropic photonic metamaterials. IEEE Journal of Selected Topics in Quantum Electronics. 2010 March;16(2):367–375.
26. Falcone F, Lopetegi T, Laso MAG, Baena JD, Bonache J, Beruete M, et al. Babinet principle applied to the design of metasurfaces and metamaterials. Phys Rev Lett. 2004 Nov;93:197401.
27. Wang Z, Yao K, Chen M, Chen H, Liu Y. Manipulating Smith-Purcell emission with Babinet metasurfaces. Phys Rev Lett. 2016 Oct;117:157401.

28. Li Z, Aydin K, Ozbay E. Determination of the effective constitutive parameters of bianisotropic metamaterials from reflection and transmission coefficients. Phys Rev E. 2009 Feb;79:026610.
29. Hasar UC, Barroso JJ, Bute M, Muratoglu A, Ertugrul M. Boundary effects on the determination of electromagnetic properties of bianisotropic metamaterials from scattering parameters. IEEE Transactions on Antennas and Propagation. 2016 Aug;64(8):3459–3469.
30. Hasar UC, Barroso JJ. Retrieval approach for determination of forward and backward wave impedances of bianisotropic metamaterials. Progress In Electromagnetics Research. 2011;112:109–124.
31. Ouchetto O, Qiu CW, Zouhdi S, Li LW, Razek A. Homogenization of 3-D periodic bianisotropic metamaterials. IEEE Transactions on Microwave Theory and Techniques. 2006 Nov;54(11):3893–3898.
32. Smith DR, Schultz S, Markoš P, Soukoulis CM. Determination of effective permittivity and permeability of metamaterials from reflection and transmission coefficients. Phys Rev B. 2002 Apr;65:195104.
33. Chen X, Grzegorczyk TM, Wu BI, Pacheco J, Kong JA. Robust method to retrieve the constitutive effective parameters of metamaterials. Phys Rev E. 2004 Jul;70:016608.
34. Tretyakov S, Nefedov I, Sihvola A, Maslovski S, Simovski C. Waves and energy in chiral nihility. Journal of Electromagnetic Waves and Applications. 2003;17(5):695–706.
35. Pendry JB. A chiral route to negative refraction. Science. 2004;306(5700):1353–1355.
36. Monzon C, Forester DW. Negative refraction and focusing of circularly polarized waves in optically active media. Phys Rev Lett. 2005 Sep;95:123904.
37. Rogacheva AV, Fedotov VA, Schwanecke AS, Zheludev NI. Giant gyrotropy due to electromagnetic-field coupling in a bilayered chiral structure. Phys Rev Lett. 2006 Oct;97:177401.
38. Plum E, Zhou J, Dong J, Fedotov VA, Koschny T, Soukoulis CM, et al. Metamaterial with negative index due to chirality. Phys Rev B. 2009 Jan;79:035407.
39. Zhou J, Dong J, Wang B, Koschny T, Kafesaki M, Soukoulis CM. Negative refractive index due to chirality. Phys Rev B. 2009 Mar;79:121104.
40. Zhang S, Park YS, Li J, Lu X, Zhang W, Zhang X. Negative refractive index in chiral metamaterials. Phys Rev Lett. 2009 Jan;102:023901.
41. Plum E, Fedotov VA, Schwanecke AS, Zheludev NI, Chen Y. Giant optical gyrotropy due to electromagnetic coupling. Applied Physics Letters. 2007;90(22):223113.

42. Decker M, Klein MW, Wegener M, Linden S. Circular dichroism of planar chiral magnetic metamaterials. Opt Lett. 2007 Apr;32(7):856–858.
43. Kuwata-Gonokami M, Saito N, Ino Y, Kauranen M, Jefimovs K, Vallius T, et al. Giant optical activity in quasi-two-dimensional planar nanostructures. Phys Rev Lett. 2005 Nov;95:227401.
44. Gansel JK, Thiel M, Rill MS, Decker M, Bade K, Saile V, et al. Gold helix photonic metamaterial as broadband circular polarizer. Science. 2009;325(5947):1513–1515.
45. Menzel C, Helgert C, Rockstuhl C, Kley EB, Tünnermann A, Pertsch T, et al. Asymmetric transmission of linearly polarized light at optical metamaterials. Phys Rev Lett. 2010 Jun;104:253902.
46. Fedotov VA, Mladyonov PL, Prosvirnin SL, Rogacheva AV, Chen Y, Zheludev NI. Asymmetric propagation of electromagnetic waves through a planar chiral structure. Phys Rev Lett. 2006 Oct;97:167401.
47. Hendry E, Carpy T, Johnston J, Popland M, Mikhaylovskiy RV, Lapthorn AJ, et al. Ultrasensitive detection and characterization of biomolecules using superchiral fields. Nature Nanotechnology. 2010;5:783–787.
48. Schäferling M, Dregely D, Hentschel M, Giessen H. Tailoring enhanced optical chirality: Design principles for chiral plasmonic nanostructures. Phys Rev X. 2012 Aug;2:031010.
49. Wang Z, Cheng F, Winsor T, Liu Y. Optical chiral metamaterials: a review of the fundamentals, fabrication methods and applications. Nanotechnology. 2016;27(41):412001.
50. Liu N, Liu H, Zhu S, Giessen H. Stereometamaterials. Nature Photonics. 2009;3:157–162.
51. Hentschel M, Schäferling M, Weiss T, Liu N, Giessen H. Three-dimensional chiral plasmonic oligomers. Nano Letters. 2012;12(5):2542–2547.
52. Helgert C, Pshenay-Severin E, Falkner M, Menzel C, Rockstuhl C, Kley EB, et al. Chiral metamaterial composed of three-dimensional plasmonic nanostructures. Nano Letters. 2011;11(10):4400–4404.
53. Cui Y, Kang L, Lan S, Rodrigues S, Cai W. Giant chiral optical response from a twisted-arc metamaterial. Nano Letters. 2014;14(2):1021–1025.
54. Menzel C, Paul T, Rockstuhl C, Pertsch T, Tretyakov S, Lederer F. Validity of effective material parameters for optical fishnet metamaterials. Phys Rev B. 2010 Jan;81:035320.
55. Simovski CR, Tretyakov SA. On effective electromagnetic parameters of artificial nanostructured magnetic materials. Photonics and Nanostructures – Fundamentals and Applications. 2010;8(4):254 – 263.

56. Jones RC. A new calculus for the treatment of optical systems I. Description and discussion of the calculus. J Opt Soc Am. 1941 Jul;31(7):488–493.
57. Menzel C, Rockstuhl C, Lederer F. Advanced Jones calculus for the classification of periodic metamaterials. Phys Rev A. 2010 Nov;82:053811.
58. Potton RJ. Reciprocity in optics. Reports on Progress in Physics. 2004;67(5):717.
59. Kaschke J, Gansel JK, Wegener M. On metamaterial circular polarizers based on metal N-helices. Opt Express. 2012 Nov;20(23):26012–26020.
60. Kaschke J, Blome M, Burger S, Wegener M. Tapered N-helical metamaterials with three-fold rotational symmetry as improved circular polarizers. Opt Express. 2014 Aug;22(17):19936–19946.
61. Kwon DH, Werner DH, Kildishev AV, Shalaev VM. Material parameter retrieval procedure for general bi-isotropic metamaterials and its application to optical chiral negative-index metamaterial design. Opt Express. 2008 Aug;16(16):11822–11829.
62. Zhao R, Koschny T, Soukoulis CM. Chiral metamaterials: Retrieval of the effective parameters with and without substrate. Opt Express. 2010 Jul;18(14):14553–14567.
63. Saba M, Turner MD, Mecke K, Gu M, Schröder-Turk GE. Group theory of circular-polarization effects in chiral photonic crystals with four-fold rotation axes applied to the eight-fold intergrowth of gyroid nets. Phys Rev B. 2013 Dec;88:245116.
64. Wang B, Zhou J, Koschny T, Kafesaki M, Soukoulis CM. Chiral metamaterials: Simulations and experiments. Journal of Optics A: Pure and Applied Optics. 2009;11(11):114003.
65. Wang Z, Jia H, Yao K, Cai W, Chen H, Liu Y. Circular dichroism metamirrors with near-perfect extinction. ACS Photonics. 2016;3(11):2096–2101.
66. Koschny T, Zhang L, Soukoulis CM. Isotropic three-dimensional left-handed metamaterials. Phys Rev B. 2005 Mar;71:121103.
67. Gay-Balmaz P, Martin OJF. Efficient isotropic magnetic resonators. Applied Physics Letters. 2002;81(5):939–941.
68. Simovski CR, He S. Frequency range and explicit expressions for negative permittivity and permeability for an isotropic medium formed by a lattice of perfectly conducting Ω particles. Physics Letters A. 2003;311(2–3):254–263.
69. Verney E, Sauviac B, Simovski CR. Isotropic metamaterial electromagnetic lens. Physics Letters A. 2004;331(3–4):244–247.

70. Baena JD, Jelinek L, Marqués R, Zehentner J. Electrically small isotropic three-dimensional magnetic resonators for metamaterial design. Applied Physics Letters. 2006;88(13):134108.
71. Holloway CL, Kuester EF, Baker-Jarvis J, Kabos P. A double negative (DNG) composite medium composed of magnetodielectric spherical particles embedded in a matrix. IEEE Transactions on Antennas and Propagation. 2003 Oct;51(10):2596–2603.
72. Vendik I, Vendik O, Kolmakov I, Odit M. Modelling of isotropic double negative media for microwave applications. Opto-Electronics Review. 2006;14(3):179.

Cambridge Elements

Emerging Theories and Technologies in Metamaterials

Tie Jun Cui
Southeast University

Tie Jun Cui is Cheung-Kong Professor and Chief Professor at Southeast University, China, and a Fellow of the IEEE. He has made significant contributions to the area of effective-medium metamaterials and spoof surface plasmon polaritons at microwave frequencies, both in new-physics verification and engineering applications. He has recently proposed digital coding, field-programmable, and information metamaterials, which extend the concept of metamaterial.

John B. Pendry
Imperial College London

Sir John Pendry is Chair in Theoretical Solid State Physics at Imperial College London, and a Fellow of the Royal Society, the Institute of Physics and the Optical Society of America. Among his many achievements are the proposal of the concepts of an 'invisibility cloak' and the invention of the transformation optics technique for the control of electromagnetic fields.

About the series

Bringing together viewpoints of leading scientists and engineers, this new series provides systematic coverage of new and emerging topics in metamaterials. It covers the theory, characterisation, design and fabrication of metamaterials in a wide expanse of areas also showcases the very latest experimental techniques and applications. This series is perfect for graduate students, researchers, and professionals with a background in physics and electrical engineering.

Cambridge Elements

Emerging Theories and Technologies in Metamaterials

Elements in the Series

Spoof Surface Plasmon Metamaterials
Paloma Arroyo Huidobro, Antonio I. Fernández-Domíguez, John B. Pendry, Luis Martín-Moreno, Francisco J. Garcia-Vidal

Metamaterials and Negative Refraction
Rujiang Li, Zuojia Wang, and Hongsheng Chen

A full series listing is available at: www.cambridge.org/EMMA

Printed in the United States
By Bookmasters